CHRISTIAN FOUNDATIONS FOR LAW AND SCIENCE

Christian Foundations for Law and Science

Believe It or Not: Paradoxes as God's Creative Forces (a Faithful Explanation for Reality)

GREGORY J. GLASER

RESOURCE *Publications* · Eugene, Oregon

CHRISTIAN FOUNDATIONS FOR LAW AND SCIENCE
Believe It or Not: Paradoxes as God's Creative Forces (a Faithful Explanation for Reality)

Resource Publications
An Imprint of Wipf and Stock Publishers
199 W. 8th Ave., Suite 3
Eugene, OR 97401

www.wipfandstock.com

PAPERBACK ISBN: 978-1-6667-4512-2
HARDCOVER ISBN: 978-1-6667-4513-9
EBOOK ISBN: 978-1-6667-4514-6

06/14/23

Bible quotations are translations of the author utilizing the Masoretic text for the Hebrew Old Testament, and the Khabouris codex for the Aramaic New Testament.

To my wife Monica

Contents

Introduction

THE LORD GOD יהוה (Yahweh)[1] and his son, the Messiah ישוע (Yahshua), teach us through paradoxes. A paradox requires two *true* statements that seem to contradict one another. For example, can God create an irresistible force and an immovable object? Yes, if they are the same thing. In this book, you can learn the key to unlock any true paradox is to imagine the two seemingly contradictory truths are *also* the same thing at their core.[2] Anything in God is possible (that is, if God allows it, then it is possible in God). See e.g., Matthew 17:20, Matthew 19:26, Mark 9:23, Mark 10:27, Mark 11:24, Job 42:2, Jeremiah 32:17, Isaiah 46:10.

I read the literal Bible text in the ancient languages of Israel (biblical Hebrew and Aramaic), and will highlight scripture confirming the Bible is consistent with paradoxes. Believe it or not, paradoxes provide Christian foundations for law and science.

Paradoxes created by God are *creative* forces. But hypocrisies (which crudely mimic paradoxes) are *destructive* forces. The Bible teaches that Satan is a lying imitator (see e.g., John 8:33,[3] Revelation 12:9). Why would

1. In the Hebrew Bible, the primary name for God is יהוה. In the Aramaic gospel, the two prominent names for God are מריא (*Marya*) and אלהא (*Alha*). See e.g., the greatest commandment in Mark 12:29.

2. When something occupies two states at once, it can be said to be paradoxical, hypocritical, polar, etc. Examples are Schrodinger's cat, electromagnetism, and wave-particle duality.

3. The ancient Aramaic codex omits John 8:1–11, so this verse is numbered John 8:33. The same Greek text is numbered John 8:44. I translate the Aramaic as follows: "You are from your father the accuser. And the lusts of your father you desire to work. He killed men from the beginning, and in truth he does not stand because the truth is not in him. When he speaks a falsehood, from himself he speaks because he lies. He is also its father."

God desire to teach us about God's own paradoxical creative power by comparison to Satan's destructive power of hypocrisy? Much of the Bible is devoted to answering this question by exposing hypocrisies in human culture and character. The Messiah particularly exposed the teachers of the law, scribes and Pharisees, as hypocrites.

This book presents an original Christian philosophy that utilizes genuine paradoxes to answer our largest questions, such as "How can God exist without being created?", "What is zero?", and "Why is God just?"

At the beginning of each chapter, you will find (1) a suggested application of the philosophy to your life, (2) a practical example for your life, and (3) a logical hypothesis you can test yourself.

I have included images throughout the book because Christian philosophy on paradoxes should be fun, or at least engaging.[4]

For better or worse, this book is fundamentally different and original, departing from many climates of opinion in modern religion. C.S. Lewis once wrote, "Now I take a very low view of 'climates of opinion'. In his own subject every man knows that all discoveries are made and all errors corrected by those who ignore the 'climate of opinion.'"[5]

KEY POINTS

The key points or hypotheses we will explore in this book:

- *Paradoxes create.* God uses paradoxes to create reality. Therefore, to understand reality, we learn to recognize key paradoxes.
- *All and nothing.* The biblical Hebrew word אל (*al*) means both "God" and "nothing."[6] Paradoxically, God himself is simultaneously the All and the Nothing, which helps explain why God is invincible. See John 1:3. In Christianity, we easily recognize God commands the

4. The genre of this book is Christian metaphysical philosophy. As we navigate some deep paradoxes together, some readers will likely add the subcategory illusionism or nominalism. I hope to avoid the subcategory labels, if possible, but that will depend on you the reader to evaluate whether this book presents a logical philosophy faithful to the literal Bible text.

5. Lewis, *The Problem of Pain*, 138.

6. This word *al* is primarily translated "God," but is also translated repeatedly in the Bible as "no, nothing." And further, the word is so extensively utilized in the Bible it is even translated as a participle such as "to, in, upon."

Everything, but we almost uniformly overlook the tremendous paradoxical power that God commands over the Nothing.[7]

- *Probability illusion.* When paradoxes compete for probability, they compete for the *perspective* of God, such that God's perspective *is* probability. Reality is a meaningful illusion or enigma in the head of God. Words like *dimension* and *matter* describe our meaningful illusion.

This should not automatically be a blasphemy to say that God is and commands the Nothing, especially given the reliance herein on Bible quotes referencing God's power over this metaphysical concept. Alternatively, if ideas in this book are blasphemous (and they are just ideas),[8] ascribe it to my failure to convey the nature of my thoughts as hypotheticals regarding the awesomeness and mysteriousness of God. And for that matter, please consider all of my writing *in progress*, a sinner's journey to honor the Most High.

I loved writing this book! One of my favorite experiences was a lively discussion on a nature walk with my friend Denis Liam Murphy, a fellow author who reimagines assumptions about life (discussed in chapter 7). As the winds roared, so did ideas.

My book is not offered as "the truth," but rather it is a philosophy for discussion. The Messiah is the truth, the life, and the way. And in John 16:12, I think he opened a door for curiosity, "Again, I have much to say to you, but you are not able to hold now."

A lot of this book involves metaphysical discussion, where I speculate on what could be true. The primary weakness of my theory in this book is that I assume Randomness as conceptual hypotheticals existing *without* being created. I make this assumption because I think logically it would require more energy to *exclude* hypotheticals than allow them. This core assumption (intuitively aligned with the principle of conservation of energy) allows me to then argue that hypothetical conceptual Randomness is literally "the Nothing" that requires no creator.

7. When the word "Nothing" is capitalized in this book it refers to the Nothing as the prime source or origin of the universe (i.e., "Hypothetical randomness is the Nothing"), as distinguished from the lowercase "nothing" which is the common word we use normally (i.e., "Bilbo had nothing in his pocket except one ring.")

8. There are only so many disclaimers one can add to a book like this without disrupting continuity in presentation of theories. I ask the reader to please pardon my writing where it appears overconfident; my goal is not hubris, but searching and testing ideas. Indeed, this book itself is like a hypothetical.

Is there any biblical basis for my assumption? John 1:3 says that "nothing" came before God. What if that statement is literally true where "nothing" is a *situation* involving Randomness? If so, perhaps hypothetical conceptual Randomness, "the Nothing," came before God. The obvious rebuttal, again from John 1:3, says God created *anything that exists*. And we have no physical evidence of a conceptual Randomness as "the Nothing" to preexist God. But a counter-rebuttal would be: hypothetical conceptual Randomness is not a *thing*, but rather is hypothetical as *the Nothing*. A second rebuttal to my theory (and more persuasive to me) comes from our practical experience that things only *appear* random to us when we lack information. Indeed, only God can answer ultimate questions; and faith is therefore necessary—especially God's faith in himself.

It can sound ridiculously arrogant to say, but if this theory is correct (or correct-ish) then it helps explain where God came from, and why paradoxes are creative in our universe. I do not intend to be arrogant, but rather curious and faithful.

SPECIAL RELEVANCE TODAY

In the twentieth century, God revealed many wonders of quantum physics to mankind.

Unfortunately, here in the twenty-first century sinful mankind is already partnering with artificial intelligence to advance quantum supercomputers and other new technologies threatening to replace man via transhumanism.[9] So we are living at a pivotal time where our healthy organic relationships with God are threatened by mad science.

Do not fear. God always saves a remnant, even after Revelation 19.

On some level this book can help Christians reorient scientific philosophy to God. I reason organic and genuinely scientific mankind will ultimately prevail over artificial intelligence because God gave natural man a covenant in the Messiah, whereas artificial intelligence does *not* have a covenant with God. Simple.

Less simple is quantum physics, which is a fundamental science because it accurately predicts the probabilistic nature of light and matter.[10]

9. Krelina, "Quantum Technology," 5.

10. This book is general enough that it should transcend different quantum theories (i.e., Copenhagen Interpretation). For a helpful apologetic supporting classical Christianity in light of quantum discoveries, see Boni, *Quantum Christian Realism*.

Quantum physics, created by God, is filled with paradoxes. This original status makes quantum physics fundamentally empowering in God but also fundamentally dangerous if misused by men.

It bears repeating that humanity needs to nurture our organic relationship with God, to keep ourselves ethically oriented at this pivotal time in technological power.

WHY ARE PARADOXES IMPORTANT?

The President of the London School of Theology wrote a book about why paradoxes are important, "What if Christianity was never meant to be simple? [The book] Paradoxology makes a bold claim: the paradoxes that seem to undermine belief are actually the heart of our vibrant faith, and that it is only by continually wrestling with them—rather than trying to pin them down or push them away—that we can really worship God, individually and together."[11]

Paradoxes have these qualities:

1. Perception based.
2. Motivating people to resolve/do something.
3. Potentially unlimited value.

These qualities are also a fair description for what we are observing right now in the universe.[12]

Wielded by God, can two conflicting paradoxes create energy from Nothing and as Nothing? Like friction (such as a mountain moving up whilst an interconnected valley is moving down), can the conflict between two paradoxes create and energize space, whereby energy is created in the

11. Kandiah, *Paradoxology*, 5.

12. See Boni, *Quantum Christian Realism*, 74–75, 77 ("God created a highly rational, highly intelligible *classical reality*, and grounded it upon a highly irrational, highly paradoxical *quantum ontology*, but did so in a way wherein the preternatural processes of *quantum reality*, though situated within the larger overall framework of *classical reality*, never once steps on the rational intelligibility that is the hallmark of classical science! . . . One might say that the cosmos, in all its material grandeur issues forth (seemingly *ex-nihilo*) from the void of quantum nothingness. However, 'quantum nothingness' is not the same as 'absolute nothingness', for the former, unlike the latter, is rife with platonic existents—including *wave-functions*. The 'quantum void' then, is more accurately a void of 'physicality.'")

very *possibility* of paradoxes competing for relative dominance/probability over one another?

If so, our world should have holographic qualities, describable by probabilities. And indeed, both are true.[13]

The classical philologist Cynthia Polsley, PhD summarizes the implications:

> "If we are part of a holographic simulation, *something lies beyond it*. Can the analogy of a holographic universe hint at what it is that lies beyond? Can it lead us back to the higher reality of heaven and spiritual things? Can it lead us back to God? Excitingly, and necessarily, the answer is *yes*. The lessons of the holographic universe are not just for physicists and scientists, nor are they restricted to a world of abstractions and mathematical modeling. The lessons of a holographic universe are for the rest of us as well. They have applications for our daily walk with God."[14]

13. There is some evidence, albeit inconclusive, suggesting we are living in a holographic universe. See e.g., Afshordi et al., "Planck Data," 5 (where university scientists claim to have found strong evidence that implies three-dimensional reality is a holographic depiction on a two-dimensional cosmic surface). See also, McDonnell, "Gateway Process," 7–8 (a famous military document suggesting our world could be a hologram).

14. Polsley, *Holographic Universe*, 17. Polsley's book places the illusion as a transient stop on the way to an eternal heaven, which she describes as real and more primal, whereas the illusion is lesser (a test of faith), such that faith is the tool to reach heaven.

In this book I present some firsts, including for example: (a) an all-consuming paradox of Nothing, and (b) a paradoxical plasma of probability. As we navigate these paradoxes, we will be utilizing a layman's understanding of quantum physics but mostly metaphysics.

ANY TWO PARADOXES CAN PROVE THE PRINCIPLE

Take this example of two paradoxes that compete for relative probability:

Paradox 1 (paradox of likelihood)

The most likely situation to our human perspective is that *something* exists and we are somehow real. The most unlikely situation to our perspective is that there was no first cause for whatever exists. Therefore, the paradox is that the most unlikely thing is actually *necessary* to the most likely thing, and vice-versa.

Discover this for yourself right now by trying to explain your existence without a first cause, *and* try to explain a first cause without a sentient perspective like you to observe it. In this exercise you can see plainly that the most unlikely thing is actually *necessary* to the most likely thing, and vice-versa.

We are going to analyze how likelihood and unlikelihood converge at both 0 and 1, showing they are the same *situation* at extremes, opposites/parallels in the middle, and probabilities throughout. A torus of light has these same qualities. In quantum computing, 0 and 1 exist simultaneously in superposition, illustrating the potential practicality of this theory and by deduction its realism. We will explore this paradox of 0 and 1 in detail throughout the book.

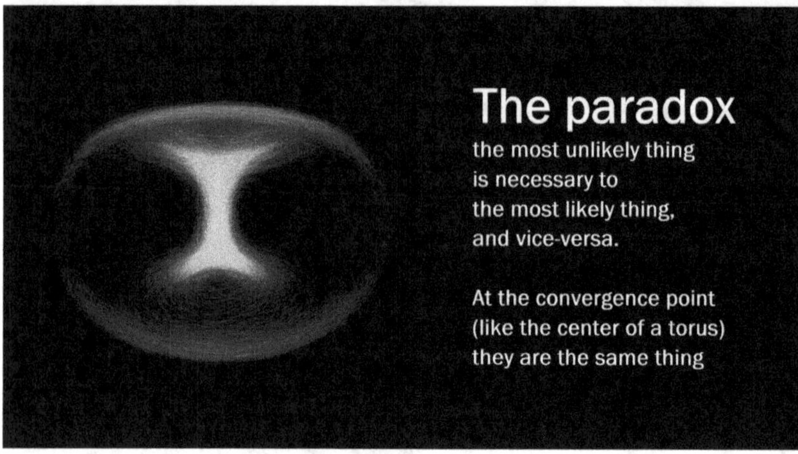

We all look different on the surface. But when you follow our lifecycle down to root elements (even beyond photons and electrons), everything is made of the same probabilistic material. Rearrange the order of materials at the root, and *anything* possible can be represented.

Paradox 2 (paradox of order)

How can there be order in oneness? The golden ratio[15] reveals order in oneness (the whole is known by knowing the relationship of the larger part to the smaller part), yet still you can observe the whole (universe) or the parts (i.e., people), but you cannot observe both *simultaneously*, lest you engage in double-counting (a math error). Either (1) you focus on the parts

15. The golden ratio is 1/1.618033 . . . It is found throughout the universe because it is the most efficient and productive use of energy. Indeed, the golden ratio is represented throughout the entire human body, as well as in nature, and in the solar system. See e.g., Meisner, "Golden Ratio." In chapter 6 we will consider golden ratio paradoxes in more detail.

individually, in which case the whole is just an abstraction or concept, or (2) you focus on the network as a whole, in which case the parts are just a perspective on the whole. But you cannot count something twice. Therefore, the paradox is that perception is simultaneously real (tangible, countable) and unreal (holographic, hypothetical).

Paradox 2 also has two points that converge. It is extremely likely that both whole and parts exist. It is extremely unlikely that neither exists, but are illusions of one another.

Paradox

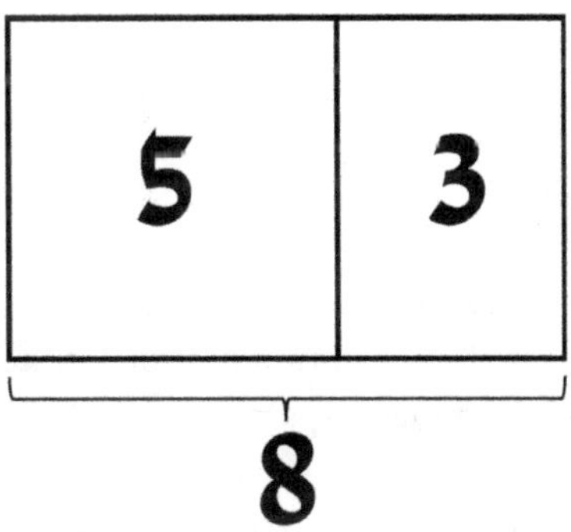

The total is 8.
Are the parts (5 and 3) real while the whole (8) is an illusion? Or vice-versa? If they are both real, the total is 16. If neither is real, the total is 0.

Like apples and cinnamon are more together than they are apart, this is only a useful perspective. The reality is the apple and cinnamon (like all energies at their core) are the same thing, and we are rewarded for growing them apart but consuming them together as one.

Ratio is holy, and in biblical Hebrew the word "sin" means to miss the mark. It is probability that makes ratio holy. So long as you keep the choice (i.e., it is either parts *or* whole) then the illusion remains functionally intact; but the moment you say it is 0 (or it is 16) then the unlikely outcome unsatisfactorily dominates the mind. People say with such confidence, "We are all one" (8) or "I am myself (5) and you are you (3)". We say these potentially contradictory statements without any sense of hypocrisy because choice between them is the implied ingredient. This example teaches us about ourselves, because we learn that choice is necessary to defending our logic.

If the solution were the 8 exists in a different dimension/plane than the parts, then this would suggest infinite dimensions, which is inefficient. But try it with only two possibilities, and the dimensions are limited to three: 0 0; 0 1; and 1 1.

A good sign of the pervasiveness of a paradox is when it can be reached and explained from multiple different methods. For example, Oxford professor Sorenson highlights another famous way to reach the same conclusion as my description of the 5 + 3 or 8 paradox, "If the power set of a set is always bigger than the set, then the universal set would be bigger than itself."[16]

By logical inference, the golden ratio proves that zero exists because of the Fibonacci sequence that defines the ratio: 0, 1, 1, 2, 3 . . . In other words, because we observe the golden ratio in all life and stars throughout the known universe, and because 0 is necessary for the golden ratio equation, therefore zero exists. This golden ratio proof is fundamentally important. It is paradoxical that the most unlikely thing (0) is proven by the most likely ratio in the universe.

New Paradox Between Two Paradoxes

Now, can we locate a paradox between these two paradoxes. How about this—observation should not be prerequisite to perception of probabilities (for observation collapses probabilities).[17] The macro-paradox shows that two things in conflict must be an illusion (0), yet also the two things are the same thing (1). The prevailing principle here is that perspective emerges from the mere possibility of an outcome. Thus, the two paradoxes are in competition to describe the universe as a probability enigma. This key point pervasively helps explain the universe—competition among paradoxes is itself the paradoxical result inherent to opposites (like 0 and 1) also being the same thing. The product of that pervasive situation *is* probability, which provides a lens through which we observe the universe. For example, quantum physics reveals all light and matter is inherently probabilistic.

Just as plants consume (grow into) light so that nature has no vacuum, the paradoxes also consume (grow into) probabilities so the entire universe is filled. Paradoxically, the imperfections of nature are perfect.

When paradoxes compete for probability, do they compete for the perspective of God, such that God's perspective *is* probability? In chapters four and six we will consider this theory by virtue of my alternate translation of the Bible's first word בראשית = "it was in his head". What if when God creates a paradox in his head, it becomes God and vice-versa, such that creation

16. Sorensen, *Brief History of Paradox*, 326.

17. Observation collapses probabilities, such that the result is a fixed outcome or choice among probabilities. In quantum mechanics, this is called wave function collapse.

and consumption are paradoxically the same phenomenon?[18] Like apples and cinnamon, paradoxes grow themselves ever appealing, to increase the likelihood of being consumed by God and God does likewise, with paradox consuming becoming God. Much of the Bible, including the supernatural example of Christ's body and blood, is about the meaning of consumption.

Readers should take time to imagine paradoxes that compete with one another. For example, my friend Scott who reviewed content for this book offered his good answers: to question #1, "Ultimate power comes from control of Nothingness and Nothingness is powerless. Nothingness can be an illusion like Everything can be—but an illusion is something, not Nothingness." And for #2, "The universe's total energy certainly is constant, while God is an uncertain amount of energy or unlimited energy, so unlimited energy would certainly destroy everything, while God is unlimited energy and created everything."

18. The network, complete in itself (like a torus), consumes itself. Consider John 6:56–57, 65, "He who eats my body and drinks my blood remains in me and I in him. As the living father has sent me, and I live myself because of the father, and he who will eat of me also he will live because of me . . . Because of this I have said to you that no man is able, that he will come to me, unless given to him from my father." To consume oneself sustainably, there must be an inversing or transference, such that the sum never collapses but rather the thing consumed actually grows and contracts the whole in fractal patterns. In this sense, you really are what you eat! Your body consumes something in order to become it—your body eats earth and then your body dies to return to earth. This is probably a useful lesson for the spirit ultimately as we ponder scalability and extra-dimensionality.

OTHER SCHOLARSHIP IN THIS AREA

Throughout this book I quote important books on paradoxes, yet after diligent search I am not aware of any previous Christian scholarship specific and detailed to my theories: God uses paradoxes to create reality, and hypothetical Randomness is source.[19]

There is substantial scholarship on the subject of "nothing" (including logical hypotheses exploring where "nothingness" is even possible).[20] And there are references throughout secular and religious literature to the subject of God and "nothing,"[21] but I am unaware of any scholarship regarding God's use of conceptual paradoxes between the All and the Nothing to constitute our universe as a meaningful illusion inside God's head.

Again, there is scholarship throughout the ages on this subject of God and nothingness generally. See e.g., well-researched excerpts from Smith, *Indescribable God*:

> "Although this is not as clearly developed, Christian theologians sometimes state or at least imply that the reason that the essence or nature is unknowable is because God does not actually have an essence or nature. The fact that there is nothing to know explains why the essence or nature of God is unknowable." Page 23 (and the same point is made by Theodoret of Cyrus, as paraphrased on page 44).
>
> "Paradoxically, it is asserted that to know God is not to know God." Page 24 (paraphrasing Cyril of Jerusalem on page 30).
>
> "[Cyril of Alexandria's] meaning is that God is not *a* being, but more than that, that God cannot even be said to *exist* since

19. But see Boni, *Quantum Christian Realism*, 122 in a different context, "why should we expect randomnities, even if stochastically constrained, to lead ultimately to persons capable of pondering the majesty of the universe? We shouldn't. That's the point. Not unless there's a Grand Mastermind at work both beneath, above, behind, and throughout the randomnity. Which is exactly what I posit!"

20. Sorensen, "Nothingness," sections 1–10.

21. See e.g., Halaw, *God is Nothingness*, 33, 66, written by a Buddhist on respecting God as Nothing, yet the author did not include any reference to randomness in the book, but writes, "Because we view everything as external objects, we do not experience the divinity in 'being'. To remedy this, treat them all as *Neti-Neti*, 'Not this, not that.' Paradoxically, this then liberates 'being' by uncovering its true nature, as a manifestation of Nothingness... Neither is Nothingness a conceptual or abstract category. It is directly apprehensible." The religious study of Tao and similar is filled with references to nothingness.

only beings exist. For this reason he calls the divine infinite and incomprehensible." Page 31 [Greek text omitted].

"Being simple would explain why the essence of God is unknowable: it does not exist to be known." Page 33 (describing Basil of Caesarea).

"Given that Pseudo-Dionysius refers to God as superessential and supernatural, it seems that the reason that God cannot be known from his own nature is that God does not have a nature. Human reason operates discursively by distinguishing one thing from another on the basis of their respective natures; but God is above all such distinctions and so cannot be known." Page 45. See also the comical observation, "Generally there is little awareness or concern on the part of Christian theologians that the concept of essence or nature is foreign to Scripture." Page 55.

"[E]mptiness itself, which characterizes all dharmas, is that which is independent and unconditioned and so truly exists . . . It would be useful for Christian theologians in certain socio-linguistic contexts to make use of the concept of emptiness in their attempt to explicate the biblical teaching of the otherness of God." Page 167.

In chapter 7 we will explore evidence that the closer we get to possibilities (via an open mind) the more we heal our bodies, because the closer we get to God the more we heal our minds. Compare a similar idea articulated by Smith, *Indescribable God*, 51, reviewing the writings of Gregory Palamas, "What can be known about God is God's energies (or energy), which describe God's effects, or how human beings experience God. In other words, God's energies are God's modes of action and interaction with created beings. Everything that can be said about God, even the term 'God,' is a statement of the divine energies."

The study here becomes even more fascinating when we start paradoxically reimagining the word "nothing" actually means something. For example, consider the sentence, "There is nothing in God that is not God."[22] What if it was funny to God (in a good way) to read the sentence two different ways: (1) where nothing means nothing, and (2) where Nothing is treated as a thing. The same could be said ironically for "God needs nothing."

Academically, Sorensen summarizes a Christian and secular history of thinkers trying to figure out what to do with paradoxes, "Augustine deployed Christian dogmas to drown the skeptical paradoxes. However, Christianity itself generates paradoxes—at least for Christians . . . God

22. Dolezal, *God without Parts*, 2.

created the whole world from nothing. In *The Timaeus*, Plato allowed that the demiurge began the universe in the sense of organizing a preexisting state of chaos. But everyone in antiquity agreed that the universe could not have had a beginning."[23]

Chaos cosmogony posits an origin of the universe from chaos. Yet among ancient Greek authors, no one seemed to ever equate their concept of *apeiron* (meaning: unlimited or original material) with fundamental hypothetical Randomness.

Reading authors like Sorensen, one gets the impression the ancients focused on other things when discussing origins (and obviously no one had the foresight to explicitly discuss quantum mechanics). Indeed, it appears from various legends (Greek, Roman, tribal, esoteric, etc.) that a predominant theme is light and darkness. The light is associated with creation and order, whereas the darkness is associated with pre-creation and chaos. As ancients observed order and intangibility in the stars at night, this influenced their philosophical perspectives. Today, we are not a society of stargazer philosophers but rather researchers and inventors with philosophies continually updated among new scientific discoveries.

In Genesis 1:2 we find unchanging foundational history, "And the earth was without form and empty. And darkness was upon the face of the deep. And the Spirit of God moved upon the face of the waters." Technically, there is no mention of a "universe" in the Hebrew or Aramaic Bible or even a separate word construct for "universe." Modern Hebrew attempted to change the word עלמ ("age/world") into "universe" but this does not work biblically (because that word עלמ and its conjugations are already used frequently and consistently to refer to limited worlds and ages). This is explained in detail in chapter 4.

It is quite telling that countless ancient thinkers made educated guesses about origins yet neglected to predict quantum mechanics, or to account for time as order as a subset of probability. Lacking these modern insights, perhaps the ancients could not equate and apply Randomness with the zero and explain their analysis scientifically. So, there are unique philosophical opportunities today in this field.

There is a modern Jewish professor, Daniel Chanan Matt, who was planning to make significant contributions in this area by writing a book about the concept that Nothing is אין, but the book about Nothing was ironically never written. Perhaps it will be someday, but in the meantime

23. Sorensen, *Brief History of Paradox*, 173–74.

Professor Matt has produced some helpful articles on the subject. From his most contemporary article[24] and original papers[25, 26]

> "'*Ayin* (Nothingness) is more existent than all the being of the world. But since it is simple, and every simple thing is complex compared with its simplicity, it is called *Ayin*.' David ben Avraham ha-Lavan (fourteenth-century Kabbalist), *Masoret ha-Berit*, ed. Gershom Scholem, *Qovets al Yad*, n.s. 1 (1936), 31." Page 1.
>
> "'When one sows a single seed, it cannot sprout and produce many seeds until its existence is nullified. Then it is raised to its root and can receive more than a single dimension of its existence. There in its root the seed itself becomes the source of many seeds.' Dov Baer, *Maggid Devarav le-Ya'aqov*, Rivka Schatz-Uffenheimer (ed.), (Jerusalem 1976), 209. See John 12:14: 'Unless a grain of wheat falls into the earth and dies, it remains alone; but if it dies, it bears much fruit.'" Page 3.
>
> "One of [the mystics'] favorite strategies is to call God 'Nothing.' We hear this paradoxical divine epithet in the East and the West: Meister Eckhart's *Nichts*, St. John of the Cross' *nada*, the Taoist *wu*, and the Buddhist *sunyata* and *mu*. I will focus here on the Jewish mystical concept of *ayin*, 'nothingness.'" Page 43.
>
> "Since God's being is incomprehensible and ineffable, the least offensive and most accurate description one can offer is, paradoxically, *nothing*." Page 43.
>
> "John Scotus Erigena, a ninth-century Christian mystic influenced by Plotinus, was perhaps the first to apply the term 'nothing' to God. Writing in Latin, he calls God *nihil*, by which he means not the lack but the transcendence of being. Because of 'the ineffable, incomprehensible and inaccessible brilliance of the divine goodness . . . it is not improperly called 'nothing' . . . Medieval Christian mystics who speak of divine nothingness, such as Meister Eckhart and Jacob Boehme, are indebted to John Scotus." Page 43.
>
> "Through negative language and paradox, these three traditions teach us that we misunderstand and misrepresent reality. We do not know the true nature of things. Modern science advances a similar critique." Pages 30–31.

Taken all together, Christian scholarship is needed to analyze the meaning of God's new and ongoing scientific revelations about paradox.

24. Matt, "Ayin," 1, 3.
25. Matt, "Ayin: Concept of Nothingness," 43.
26. Matt, "Varieties," 30–31.

Chapter 1

The Grand Paradox

SUMMARY

God is paradoxically the All and the Nothing. How is this possible? It is theoretically possible mathematically, that reality is an illusion created by God (or inside God's head) from *hypothetical* Randomness. To posit that hypothetical Randomness is self-emergent and requires no creator, I deduce it must be effortless for the universe to simply exist effortlessly as hypothetical possibilities, because it would require energy to *exclude* hypothetical possibilities.

Relevance for your life . . .	Imagine for your life . . .
You are made in God's image. You want to understand your relationship with God. When you learn God works in creative paradoxes, you learn to see your own creative paradoxical behavior in life, such as *giving is receiving*. And you learn to exclude Satan's copycats—hypocrisies—that are not genuine paradoxes.	You are made in God's image. Your thoughts feel like who you are, yet they are also intangible like hypotheticals.

Logical Hypothesis

1. If Randomness = Nothing,
2. Then God emerges with his will in respect of Randomness and God,
3. Therefore reality is necessary either way.

GRAND PARADOX: GOD IS SIMULTANEOUSLY THE ALL AND THE NOTHING

The Lord God יהוה is simultaneously the All and the Nothing. How is this possible? God can use his quality of Nothingness to be invincible. What is more invincible than Nothing? It is so invincible you cannot even touch it, and attempts to identify it are mere hypotheticals. Just as the All is in everything to create the probability that we exist, so too the Nothing is in everything to create the probability that reality is a grand illusion (a holographic abstraction).

My dictionary at home defines an illusion as "perception of something objectively existing in such a way as to cause or permit misinterpretation of its actual nature either because of the ambiguous qualities of the thing perceived or because of the personal characteristics of the one perceiving or because of both factors."[1]

In Strong's Hebrew Dictionary, number 6754 (צלם), the word "illusion" is synonymous with "image," as we read "the word means image in the sense of essential nature".[2] Remember Genesis 1:26, "Let us make man in צלמנו (our image)." See also Genesis 5:3. In other words, images are like shadows of reality. And see Psalm 139:12 referring to God, "Also the darkness is not dark from you, and the night shines like the day, as the darkness is like the light." Psalm 18:11 also referring to God, "He made darkness his secret." Isaiah 43:10, "Before me no god was formed, nor after me shall there be." See also the Hebrew scholar d'Olivet (צל is "void itself").[3]

The notion of reality as an illusion is well studied. One of the more famous quotes is attributed to Albert Einstein, "Reality is merely an illusion,

1. Gove, *Webster's Dictionary*, 1127.
2. Strong, *Hebrew Dictionary*, 238.
3. D'Olivet, *Hebraic Tongue*, 434.

albeit a very persistent one."[4] Further, Einstein's "[g]eneral relativity demands that space-time, the four-dimensional structure that space and time together form, is smooth and continuous, whereas a quantum description requires a pixelated description. As the theoretical physicist Nima Arkani-Hamed has said: 'Almost all of us believe that space-time doesn't exist, that space-time is doomed, and has to be replaced by some more primitive building blocks.' Admittedly, no one yet knows what those might be—but our insights suggest the hunch they must be replaced is right.'")[5]

There is an inherent oneness between faith and the Nothing—in other words, sometimes faith looks like nothing, and sometimes nothing looks like faith.[6] Can you put your finger on faith? You cannot. You can put your finger on processes, even neural process, that result in faith. But faith is an intangible description. The Nothing is an intangible description.

We use All and Nothing as such causal words (adjectives) that they become paradoxically ubiquitous, paradoxical because they are (1) never actually observed and/or (2) observed so pervasively in every moment that observation itself is completely dependent upon them. Some people will feel we are quite obviously living in a paradox. Others find the notion ludicrous. Probably most people just have an open mind to the possibility that God uses paradoxes in some way.

4. Guynup, "Is Seeing Believing," 1.

5. Hoffman, "Is Reality Real," para. 24.

And to further emphasize this subject must be addressed by Christians, be aware of the atheistic/agnostic perspective, such as Daniel Dennett who writes and attacks a lot, "you can't be a satisfied, successful illusionist until you have provided the details of how the brain manages to create the illusion of phenomenality, and that is a daunting task largely in the future. As philosophers, our one contribution at this point can only be schematic: to help the scientists avoid asking the wrong questions, and sketching the possible alternatives, given what we now know, and motivating them—as best we can." Dennett, "Illusionism," 66.

6. An insightful pop culture reference comes from the film The Neverending Story, where the antagonist is a force called "the Nothing," which destroys the world of Fantastica. The plot twist (and paradox) is that the Nothing prevails only when people stop believing in fantasy.

To help explain creation, ask this: how could raw fundamental hypothetical Possibility create a ripple in Nothingness? How could Nothing become lopsided? How could a signal emerge within Nothingness? What is more primal than a physical thing, spiritual thing, and conceptual thing? Does Possibility become Nothing in order to consume it, and vice-versa?

Faith is the only tool small enough (wave length = 0) to interact with the Nothing. Would the Nothing and faith interact as paradoxes to create an illusion, to create reality? And what is the meaning of law and morality in an illusory world? This book addresses these questions. Perhaps the illusion of reality protects or shields that which is *truly* pure, an ethereal Godly realm. Meanwhile what if the illusion we know as the universe creates a test (on the surface) to see who is worthy to enter the deeper structure within?

What if, at a fundamental level of universal creation, the exclusion/indifference to possibility is a possibility itself where paradoxes are inherent to the abstraction? Can the Nothing be paradoxically both conceptual and real? If so, are concepts and reality the same thing? What if reality is thought, God's thought?

These are the kinds of mostly rhetorical questions this book frequently presents. You might tire quickly if you do not stop and apply the ideas to your personal perspective. Remember, the beginning of each chapter has text box suggestions to help you apply the macro information to your own life.

SIMPLEST SOLUTION

The simplest solution may be the greatest truth. What if this reality you are experiencing right now *is* what the Nothing looks like?! And is God's faith both nothing and everything simultaneously? Because our earth looks and feels real, it seems exponentially unlikely that this reality/universe is (also) what the Nothing looks like. But what if that is precisely why probability manifests within Nothingness? What if the tension between likelihoods is the abstraction that is reality? What if reality, like God, is both the Nothing and the Everything?

John 1:3 is instructive. The verse is not translated easily into later languages like English, but in the ancient Aramaic language spoken by the Messiah, the preserved Aramaic text (called the Peshitta) provides the possibility that God is both the Nothing and the Everything because the gospel says so literally. I read this language well and can write a hyper literal translation into English to emphasize this point, "Anything in his hand existed and without him not even one existed whatever that was." The phrase "not even one" means "nothing" in Aramaic.

This is a good teaching example of God's character shown through both the Hebrew and Aramaic languages. Because the word for "every" (כל) also means "any," when Jewish people wanted to emphasize the word "nothing" in comparison to "everything," they would also state the negative ("nothing"). For example, in Hebrew if you wanted to say "I picked up every apple," you would actually say "I picked up *any* apple and not one remained". The reason for the second part ("not one remained") is that "any" in Hebrew can also mean "every" in context, so it is inherently ambiguous by itself (meaning ambiguous out of context). This example likely reveals a great truth about the nature of God. He teaches us. Earth is our context.

BIBLICAL HEBREW

In biblical Hebrew, there are many words for "nothing" and related terms such as no, never, lacking, and empty. Ironically, the word nothing means something. And in many verses, that meaning is very special because it describes God's dominion and power. For example, Job 26:7 regarding God, "He spreads out the north upon emptiness; he suspends the earth on *nothing*."

Here is the big one: the biblical Hebrew word אל means both "God" and "nothing."

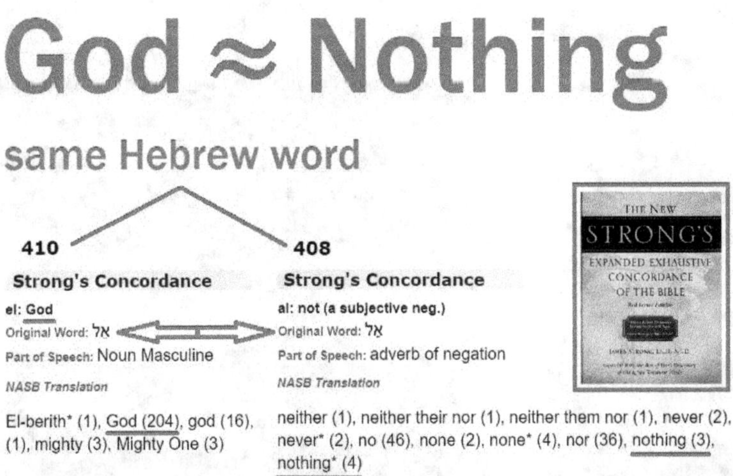

The greater meaning is that God is the ultimate lawgiver with the power to say yes *and no*. God can create anything, and bring anything to complete and utter annihilation.[7] Perhaps in God's sense of humor, nihilists have always been a little bit correct, but for completely the wrong reasons.

People often refer to God using opposite terms (positive and negative), such as creator and destroyer; knowable and unknowable; nameable and un-nameable; natural and supernatural; like and unlike. In every opposite, there is potential for paradox because God is big enough for unlimited

7. Many Bible verses describe God's ability to annihilate and utterly extinguish anything God chooses. See examples such as Matthew 7:13, 10:28, and 21:41–44, Revelation 20:14, Ecclesiastes 3:20 and 12:7, and Isaiah 26:14.

paradoxes. In this way, what if we honor God by recognizing God's paradoxical power? I think so.

Possibility has the same quality that the result is unknown or unlimited. For example, is science simple, complex, or exponential possibilities in between? Anything in God is possible, meaning that so long as something is in God, then it is possible. Matthew 19:26. The only negative that cannot fit God is "*im*possible," unless the atheists are right that in some way God is also impossible—and it would actually be quite funny and interesting if God proved the atheists right in this manner!

As we write and talk about God, eventually we use enough esoteric words that we can pretty much ascribe any ethereal concept to God: all, nothing, dimension, space, time, inside, outside, anything. And the volumes written by God's unauthorized biographers (myself included) could half-fill a half-empty cosmic swimming pool.

There are abundant reasons to be cynical when studying paradoxes. For example, much paradoxical thinking/writing involves identifying puzzles created by the way we use words,[8] and then defending our approach to the user of words, and if the defense is successful (or not) we say "paradox." We congratulate ourselves more if the defense is successful. This creates a bias in favor of simple paradoxes rather than paradoxes built on other paradoxes or in competition.

In my view, the best paradoxes (the real ones), should be revealed as truths by harmony with the literal Bible.

GOD IS

In regards to The Nothing and The One, imagine if God is the only One who can say, "I have no one." In this, God is both paradoxically alone and not alone. God would have the Nothing (with all its attendant qualities, such as "no one"). People agree, "there is no one like God," and yet the Nothing can be theorized to share qualities with God, because God controls and is the Nothing just as surely as God controls and is the Everything (the One). At least, that is the theory here.

It would be very interesting indeed if God solved the loneliness problem (also an accountability problem) by the quality of inherent companionship with the Nothing.

It is a useful exercise to apply each role of God paradoxically to One and Nothing. For example, God the Father: the Messiah Yahshua had no earthly Father, yet was one with the Father in heaven. God the Word: in the beginning was the Word, the first, yet the text of John 1:3 says before God there was "nothing", so a potential deduction is that the Word comes from Nothing. Or, the One comes from zero. By having faith in One, perhaps you concurrently have faith in zero, because they are the same thing (as will be considered with detailed examples later in this book).

Opposites are sometimes designed to be in harmony, such as father and mother. They unite as one, and the product of them (child) *is* them.

8. "According to [Ludwig] Wittgenstein, most paradoxes can be nipped in the bud by bringing words back into their natural settings, by studying how they are taught to children, and by noting the role they play in larger practices." Sorensen, *Brief History of Paradox*, 336.

The nihilist would counter that a hypothetical Nothing would have no qualities whatsoever to share with God or anyone else. But what if the nihilist is asking a misdirected question about thinginess? What if, within fundamental hypothetical Randomness, the qualities of Nothingness (such as emptiness, intangibility, randomness) are inherently hypothetical such that it would *require* energy to exclude a hypothetical? And if one hypothetical can effortlessly exist within Randomness, why not many? The nihilists do not know, and neither do I. Perhaps the question then becomes whether we like this trait about ourselves (unknowingness), and what we do with it.

I love the idea God is so great that the Nothing and God would share the quality of belief or faith. What is belief, if not nothing? It is empty, intangible, elusive. And yet it is creative and empowering. Which is it? What if the tension of the paradox creates the energy that we humans call possibility?

There is also danger here in navigating paradoxes. A synonym for paradox is "contradiction," which is often discussed by the famous nineteenth century philosopher Georg Hegel. He found them to be creative, as he wrote, "He who claims that nothing exists which carries in it a contradiction as an identity of opposed determinations is at the same time claiming that nothing alive exists. Indeed the force of life and, even more, the power of the Spirit, consists in positing the contradiction in itself, in enduring and overcoming it . . . contradiction is the root of all movement and vitality."[9] Compare the thirteenth century philosopher Thomas Aquinas who believed that contradictions could not come from God, "Nothing which implies contradiction falls under the omnipotence of God."[10] Perhaps if Aquinas changed his perspective on "Nothing" he would still find his statement to be true. In any case, the danger is noted, especially as Satan is an imposter and a hypocrite.

LOGICAL HYPOTHESIS

Recall the logical hypothesis at the beginning of the chapter.

1. If Randomness = Nothing,[11] then
2. God emerges with his will in respect of Randomness and God, therefore
3. Reality is necessary either way

And the meaning of this logical hypothesis provides a staggering example of God's love. I will explain why in the next several paragraphs. Please bear with me.

If we assume a primordial plasma of infinite Randomness (i.e., the Nothing), then simultaneously or eventually (after enough iterations of Randomness) *anything* can manifest, including a self-aware being (namely, God). Atheists would argue this is the same classic thought experiment of the infinite monkey theorem—given infinite time, monkeys on typewriters will eventually and randomly type out Shakespeare's entire volume of works.

9. Sorensen, *Brief History of Paradox*, 307–8.

10. Lewis, *Problem of Pain*, 16, quoting Aquinas, *Summ. Theol.*, I Q XXV, Art 4.

11. As we will explore in chapter 2, absolute and total Randomness is noncausality. Absolute and total Uniformity (certainty) is also a form of the Nothing because there is no way to distinguish or discern anything from anything else. In total Randomness you learn nothing because logical progression is impossible to discern. In total Uniformity, you observe nothing because there is no contrast to stimulate brain cells.

But when I write of Randomness, I mean the randomness of *concepts or hypotheticals*, meaning many concepts interacting and existing in a completely random and nonsensical manner.[12] (The materialism later borne from concepts (much later), such as Earth, will follow the code (will) of the Creator, but we have not reached that point yet in this brief analysis; we are still in a primordial plasma.)

In a primordial plasma of Randomness, the moment that God emerges is the same moment God becomes aware of Self. How could God emerge from Randomness if the Bible states God is primal? See e.g., Isaiah 43:10, states "Before me no god was formed, nor after me shall there be"; and John 1:2, "without him not even one existed whatever that was." The answer would be that these verses (read literally) do not preclude Randomness from preexisting God, especially if the logical hypothesis is true that Randomness = Nothing (because then these Bible verses gain new meaning/perspective, to say quite literally what people already know, "Nothing came before God"). One could say God is fundamentally humble because he comes from Nothing.

Moreover, the moment God emerges from a conceptual Randomness plasma, God would recognize he is in a race or contest with Randomness. Theoretically, God would see his job to permeate Randomness and prevent

12. Principles are timeless. It defies the human mind trying to imagine prior infinity of time or space or matter. But if you imagine an abstract principle (like 0+1=1) simply being possible in a timeless hypothetical randomness, that is workable.

the unjust from randomly occurring (i.e., to prevent the infinite monkeys from typing something torturous).

However, if Randomness is infinite, then get ready to be scared because Randomness would have already manifested unjust torture before God could stop it, unless (get ready to be comforted) God's essence is so simple that statistically God would manifest from Randomness *before* anything complex would manifest (such as torture). If this assumption of relative likeliness proves true (and it emphasizes the importance of probability to universal justice), then Randomness was never afforded enough godless iterations to manifest unjust torture.

And now for the loving realization: this service that God provides to all of reality as Top Cop (the One who stops the infinite monkeys from their random typing of awkward forms of torture), is beyond words for the gratitude it deserves from us.[13]

God's pattern recognition is his original divine aptitude in the primordial sea of Randomness, effected by his further aptitude (will) to propagate *desired* ratios such as phi.

Why would God's will emerge from Randomness? Because it is a principle of Randomness that given exponential iterations, eventually coherence is manifested (such as fractal form). The only capacity needed from there (upon the emergence of a fractal form that reflects on itself) is the ability to harness the coherence without self-destruction.[14] This will be explained in chapter 6. When you think about this deeply, it emphasizes the sacredness of God's memory.

And so, we arrive at the paradox: It requires will to prevent the formation of will from Randomness, which evidences will is self-emergent and primal, because it exists *either way*.

This means that God exists because God chose to exist. He could have allowed the Randomness (in which he emerged) to consume his self-reflecting fractal quality, but instead he chose to become God and therefore to become the Hero who stops Randomness from iterating into unspeakable torture.

13. If God was faithless, he would dissolve back into Randomness, and it is frightening to speculate the countless iterations of formation and dissolution in a universe without our faithful God. This concept was even featured in the secular film The Matrix Reloaded, where Neo meets the Architect and questions how many iterations of the program he has *already* been through.

14. DNA is a real-world example of a fractal code that does not self-destruct, but perpetuates itself through reproduction.

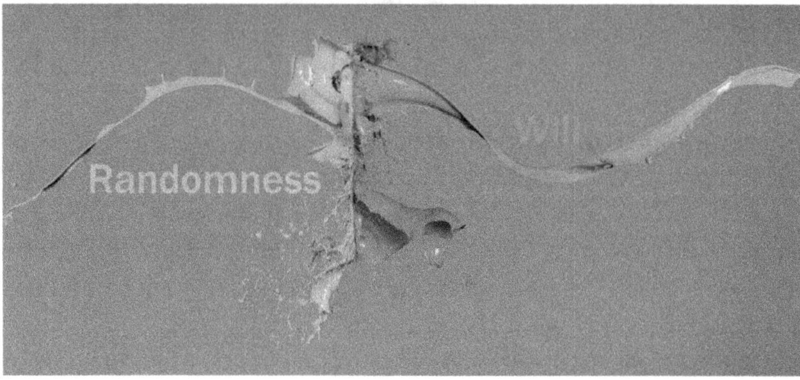

God is truly our Creator and our Savior. Imagine the tools he has developed to quiet and harness Randomness. Our Hero has so permeated the Nothing that he is One with it. What if Randomness even serves as some ongoing source of energy for God?

If God ever chose to dissolve Himself, reality would divert back to the plasma of conceptual Randomness. Praise יהוה for His strength, patience, endurance, and love. Mark 12:29–30, "The first from any of the commandments, 'Hear Israel, the Lord our God is one Lord, and that you love the Lord your God with any of your heart and with any of your soul and with any of your mind and with any of your strength.' This is the first commandment."

Regarding using a paradox to honor God, the Christian author Rocco Boni said it well, "This gives way to a paradox of sorts. Quantum outcomes do obtain, *but without sufficient causal precedent*. Quantum event outcomes are thus deemed a-causal. *Yet they do occur!* This gives rise to the question: If quantum event outcomes are not determined by anything within the universe, could quantum event outcomes be sufficiently determined by something outside of, or beyond the universe? . . . As far as this physical universe is concerned, all is being continuously *willed* into physical existence (again, constituent-particle-by-constituent-particle) by what seems to be a transcendent volition—a preferential 'Will' so to speak."[15] In summary, Boni is emphasizing that God's will is primal in order to choose among the exponential possibilities. Recall Revelation 4:11 regarding God, "Because of you anything was created. And by way of your will, being was created."

15. Boni, *Quantum Christian Realism*, 12.

LOGIC BY ANALOGY

What if God is the ultimate paradox? In Hebrew, God's name is אשר אהיה
אהיה ("I am who I am"), which is an inherently circular mystery. A paradox is also an inherently circular mystery.

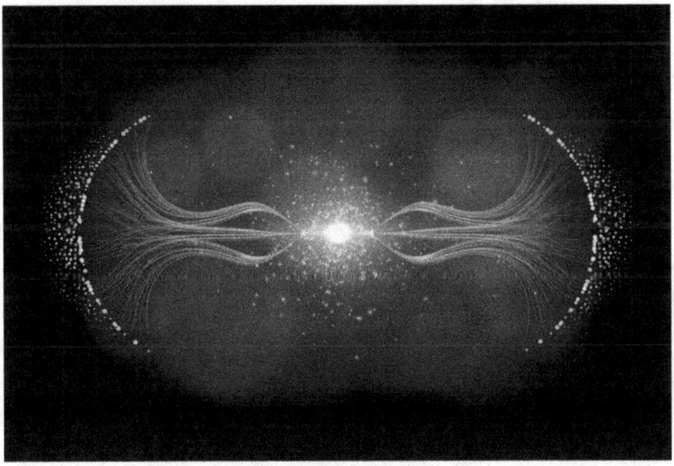

Try to imagine a better way to conceal the Nothing than as an illusion of anything (All). Both All and Nothing are omnipotent in their own realm, and in different ways (such that Nothing is omnipotent by touching nothing and being untouchable; whereas All is omnipotent by being everything and everywhere). One might say, the All and the Nothing respect one another as mutually dependent illusions, perhaps even consuming one another.

Consider these logical hypotheses:

1. If no possibility of nothing, then anything.
2. If possibility of nothing, then possibility.
3. If no possibility of anything, then nothing.

Is the possibility of possibility a self-emergent concept? The nihilist would ask, "Why should a concept or possibility exist within the Nothing?" Consider several possible answers.

First, the Nothing is paradoxical, meaning a hypothetical conceptual illusion. An illusion is nothing. A hypothetical is nothing. A concept is nothing.

Second, the Nothing is just as likely as anything, so Nothing resides as pure Randomness which is the *absence* of likelihood. So, the key to understanding the place of Nothing is to consider it with abstract mathematical lenses.[16]

From a mathematical perspective, the Nothing really would be synonymous with fundamental hypothetical Randomness, because Randomness excludes the only possibility there ever was: probability. In order to exclude Everything, Nothing must eradicate possibility itself, which is impossible; hence Nothing is impossible, which is paradoxical. The nihilist would say (in part correctly) this is mere wordplay.

But there is something here to develop philosophically. What if in the tension or paradox between conceptual Randomness and conceptual Probability, we experience reality, the grand illusion between concepts where it requires more energy to *exclude* something than allow it, like the ease of filling a vacuum in nature rather than maintaining it? If so, Occam's razor (or conservation of energy), the result is that hypotheticals exist because they are easier to allow than they are to exclude. That one fact would be a game changer for creation philosophy.

16. "Of course, the question of exactly *how* a quantum wave-function, which is literally an abstract entity, can possibly engage with, and/or interact with physical reality is a fundamental (and ontological) mystery of quantum mechanics unto which there is no answer (nor can there ever be one). All we know for sure is that quantum waves (which are non-physical) become quantum particles (which are physical), and quantum particles (which are physical) become quantum waves (which are non-physical)." Boni, *Quantum Christian Realism*, 9.

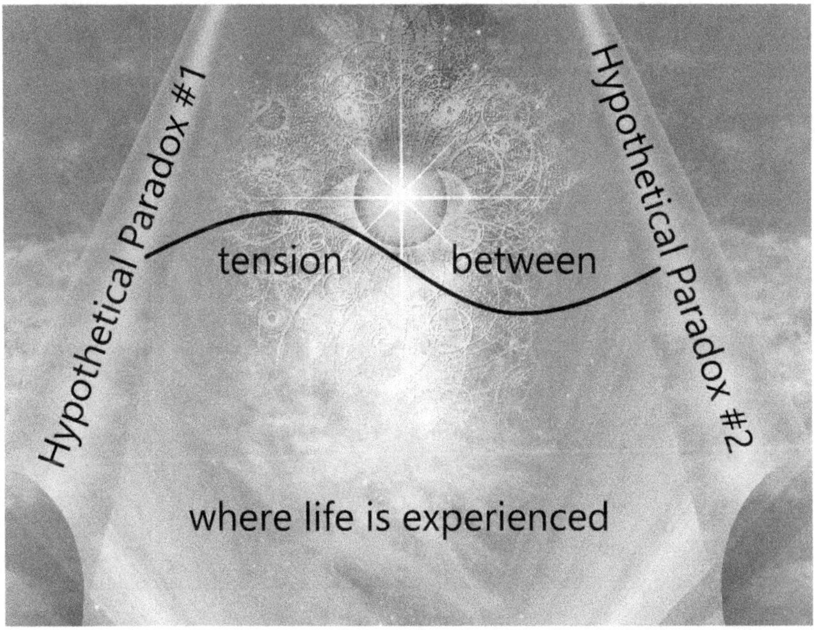

In this sense, the Nothing would be a description of Randomness, not a description of zero (absence of anything). And the reason for this is Randomness itself, which is an inevitable or fundamental quality, an axiom, that the Nothing is just as likely as probability. One could imagine that Randomness reflects upon itself, and in the displacement of that reflection the opportunity arises for Probability to manifest. And vice-versa, in the reflection of Probability, Randomness can manifest.

Consider this paradox framework explained simply, "A negative existential is a statement that denies the existence of something. How can such a statement be true given that there must be something for the statement to be about?"[17] Give it a try. Use your words or thoughts to deny the existence of Nothing, and you quickly find yourself in a paradox. Apply the logic: only the Nothing could deny itself, and in doing so would validate itself as Nothing. Once again, the paradox shows unity at the root (denial is nothing). This is like saying only God can use God as a verb, to be Godding. Only Nothing can Nothing. Can God destroy Nothing?

17. Sorensen, *Brief History of Paradox*, 29.

"According to [Thomas] Aquinas. . . God is impassible; nothing can cause God to do anything. This is guaranteed by God being fully realized; whichever property he can possibly have, he actually has."[18]

At this point in the book, we are thinking about ways that God and Nothing and Randomness potentially share qualities. For example, being "beyond comprehension" is a quality they share. You cannot even answer questions about God's power and existence without using the word "nothing" or its synonyms. For example, what came before God? Who is like God? What does God need?

In this theory, the opposite of God's will is fundamental Randomness (which is Nothing). Paradoxically, the reason that God's will is primal is that it is necessary to counterbalance the abstraction of the Nothing as Randomness.[19]

And how does displacement accomplish a creative force? It does not. Rather, the displacement reveals the two forces were always one and the same: Randomness and Probability.

Whereas spatial orientation to humans refers to order along dimensions (i.e., physical things entering empty space), it is not so for a purely

18. Sorensen, *Brief History of Paradox*, 180.

19. Important scholarship exists highlighting the primacy of God's will. In addition to Boni, see e.g., Dolezal, *God without Parts*, 165 ("In all his willing he is his own ultimate object of desire, as it were. Furthermore, the act by which God wills himself is identical with the act by which he exists. His action of volition is not something in him in addition to his very essence. The will of God is simply God willing.")

mathematical hypothetical conceptual perspective of Randomness. There is no *entering* in the Randomness hypothetical, but rather empty space is a mere hypothetical possibility in the tension between Randomness and Possibility. If the Randomness was not inherently present, then there would be no space to describe it conceptually. So, space enters Randomness, not vice-versa. Randomness does not *enter* anything, for Randomness *is* anything—that is the paradox, because anything is supposed to be probability, which is supposed to be the opposite of randomness! The two opposites once again appear to be the same thing.

C.S. Lewis wrote that the medium (space) that separates us also brings us together.[20] This is a paradox. If there is no space then all is one, and we would know zero distinction between self and non-self, thereby causing zero growth, which is death (zero). But if there is space, then we have a common medium to send messages and grow. As the number of possible relationships grows (by increasing all factors such as distance, time, space, matter) possibility itself manifests. As explained in chapter 6, ratio is holy, and God is allowing us to interact with ratio and be ratio.

In a world of illusion, some would say ratio does not *describe* the thing, but rather ratio *is* the thing. Whether correct or not, it emphasizes a logical deduction from Lewis' observation that the universe must allow matter to separate things.

Hypotheticals allow you to measure otherwise unmeasurable things, like equations that find the angle of a triangle by assuming hypothetically an adjacent triangle. In Aramaic the word Messiah also means measure. Do we assume or believe ourselves into relationship with Messiah?

If you give a nihilist arguments like these and engage them on paradoxes, imagine the progress they could make. As quantum physicist Niels Bohr stated, "How wonderful that we have met with a paradox. Now we have some hope of making progress."[21] And again, Niels Bohr writing to Albert Einstein at the Fifth Solvay Conference on Quantum Mechanics in 1927, "Everything we call real is made of things that cannot be regarded as real."[22]

20. Lewis, *Problem of Pain*, 21–22.

21. Graham-Smith et al., *Optics*, 1.

22. Boni, *Quantum Christian Realism*, 5.

PRACTICE YOUR PARADOX SKILLS

This book can help you learn to practice your skills in identifying paradoxes and solving them. Consider the classic paradox: irresistible force meets immovable object. These are the options to resolve the paradox:

1. Both exist and are therefore the same thing
2. Both exist in a universe that allows paradoxes
3. Neither exist
4. Only the irresistible force exists
5. Only the immovable object exists

Regarding option one, an immovable object would need to be omnipresent and omnipotent, or else be literally nothing. Likewise, an irresistible force would need to acquire mass equivalent to 100 percent of all possible mass (hence omnipresence and omnipotence),[23] or else have zero mass in the sense of being literally nothing.[24] Hence, the irresistible force and immovable object would be the same thing (God himself), resolving the paradox.

And they could also be perfect opposites to resolve the paradox: if the irresistible force was everything, the immovable object would be nothing. And vice-versa also works. God solves these paradox solutions as well, for only God can be simultaneously Everything and Nothing. God is simultaneously the irresistible force and the immovable object. So, the real answer is another question: can God create himself?

23. One might wonder about object density, such as an omnipresent object that begins with uniform density but then allows energy transference within itself to remove the uniformity. That object would not qualify to the extent that internal transference is an attribute of movability.

24. If the claimed irresistible force or immovable object was only 99.9 percent omnipresent or omnipotent, the nature of quantum mechanics shows the 0.1 percent would still resist or move the 99.9 percent in some way at the quantum scale.

Chapter 2

Anything in God Is Possible, Including Paradoxes

SUMMARY

WHATEVER IS IN GOD is possible. Paradoxes are creative forces from God. But hypocrisies (which crudely mimic paradoxes) are destructive forces from Satan. The Bible teaches that Satan is a lying imitator. Why would God desire to teach us about God's paradoxical creative power by comparison to Satan's hypocritical destructive power? Even the title Messiah (משיחא) in the language spoken by Yahshua is an obvious wordplay for "possible, probable" (משכחנא).

Relevance for your life . . .

Meditating upon God's paradoxes, you can learn about the Messiah, such as his paradoxical power to judge sin by releasing sin.

You learn physical things (i.e., water, plants, rocks) are paradoxical dualities, empowering your choices on earth. For example, increase the dosage of a plant remedy and it can become a weapon, or reduce the dosage of a plant toxin and it can become a remedy. God is teaching us about ratio.

Imagine for your life . . .

People find they attract who they are.

So, when you change your mind, the energy of the whole environment changes. When you push down one side of a measuring scale, the other side instantly and simultaneously rises.

If you judge, you will be judged. Whatever you measure is measured back to you.

Logical Hypothesis

1. If anything,
2. Then possibility,
3. Then 0 to 1

PROBABILITY IS ESSENTIAL

The title Messiah (משיחא) in the language spoken by Yahshua is an obvious wordplay for "possible, probable" (משכחנא).

In a universe where all light and matter are probabilistic, it is reasonable to hypothesize God created a plasma of probability (remember the wordplay with Messiah) at the core.[1] A probability plasma would present a fascinating and paradoxical duality/mirror with a self-existent hypothetical Randomness[2] from which God theoretically co-emerged.

1. Re probability at the root of the universe, "A quantum wave-function then, is no less than a potentiality catalogue marking all the possible and/or potential states of being and/or existence that a quantum particle or system might assume upon its physical actualization (which, recall occurs when the wave-function once again engages and/or interacts with physical reality)." Boni, *Quantum Christian Realism*, 8.

2. One might wonder why a hypothetical Randomness would be primary, rather than a hypothetical of a hypothetical, or a hypothetical of a hypothetical of a hypothetical, etc. The answer is a single hypothetical is primary, but a hypothetical of a hypothetical is a derivative product that is dependent upon its source.

If you envision the smallest quantum world as a plasma of waves, where uncertainty defines the laws of relativity, the beginning and end of any given wave could be indeterminable from the perspective of another wave. So, only the whole (every component part) of this quantum world could produce the perspective to know itself. And therefore, Oneness would be a natural desire or trait of waves. By contrast in our larger material world, every component part is large enough (at least the width of a photon) to measure (interact with) others to create a constantly relative experience for the individual. This focus at the macro/material world collapses waves and therefore appears to compete with Oneness.

What if we found at the quantum level (in this plasma of probability waves) that causality and reverse-causality are indistinguishable, such that fundamental consciousness would be fairly defined as the attempt to observe the motion of plasma energy in a particular order? The absence of random patterns would be the presence of causality, and the tension between these two (Order and Randomness) would be fundamental consciousness. Even a slight pattern occurring inside the plasma (or on its boundary) would magnify exponentially into the outer structure (perhaps a torus) in which we experience classical reality. I think that kind of theory would lead toward an interesting definition of consciousness.

Consider this logical hypothesis: to say there was never a possibility would require certainty. To abstain from the question leaves possibility open. If there was always a possibility, then perhaps possibility is the first cause.

The illusion in the hypothesis is that Possibility and Nothing are mutually exclusive, when in fact they are the same thing.[3] Possibility covers all things but Nothing defines itself. It may be the tension between these two create a kind of experience through the plasma of probability, which later manifests to our human experience as a first cause—competition for relative probability to the perspective of a holographic observer.

As you read these questions, it may feel counterintuitive to think of Nothing as fundamentally necessary to the possibility of something, but even practical mainstream quantum physics has become accustomed to counterintuitive thinking. For example, here is the famous theoretical physicist Richard Feynman:

> "Throughout these lectures I have delighted in showing you that the price of gaining such an accurate theory has been the erosion of our common sense. We must accept some very bizarre behavior: the amplification and suppression of probabilities, light reflecting from all parts of a mirror, light travelling in paths other than a straight line, photons going faster or slower than the conventional speed of light, electrons going backwards in time, photons suddenly disintegrating into a positron-electron pair, and so on. That we must do, in order to appreciate what Nature is really doing underneath nearly

3. The paradox simply stated is that reality is an illusion, because like all paradoxes, the pieces are ultimately the same thing, which is discerned when they converge at their root (in this case, possibility). Cold is not a thing in itself, but rather cold is the *absence* of heat. This textbook example helps explain how opposites can indeed be recognized as the same thing. Electricity and magnetism is another textbook example, a single unified phenomenon called electromagnetism that can produce opposite effects in its component parts.

> all the phenomena we see in the world. With the exception of technical details of polarization, I have described to you the framework by which we understand all these phenomena. We draw *amplitudes* for every way an event can happen and add them when we would have expected to add probabilities under ordinary circumstances; we multiply amplitudes when we would have expected to multiply probabilities. Thinking of everything in terms of amplitudes may cause difficulties at first because of their abstraction, but after a while, one gets used to this strange language."[4]

And consider Thomas Reid's philosophy as paraphrased by Sorensen, "Paradoxes illuminate common sense by provoking bits of [induction] into consciousness. As more paradoxes are discovered, more of common sense becomes visible. Without a provocateur, common sense is faceless."[5]

A very good paradox is sustainable *because* it feeds itself. Much of the Bible is about this topic, such as John 6:53 foreshadowing the last supper at Passover, "Truly, truly, I say to you that unless you eat the body of the Son of Man and you drink his blood, you have no life in yourselves."

The best way to avoid collapsing on yourself is to *be* the very thing you interact with, including your own opposite. The Messiah goes from life to death to life. Taken to its fullest or most extreme example: Everything *is* Nothing *is* Everything. And poetically, this can feel like life is a fractal illusion, which emphasizes why the revelations of holographic science are so important.

Here is another example of opposites potentially being the same thing: how can contraction *be* expansion? If vortex-based mathematics is correct then 0 and 9 are the same number, such that as you expand away from 9 (i.e., going from 7 to 6) you simultaneously contract toward zero.[6] By

4. Feynman, *QED*, 119.

5. Sorensen, *Brief History of Paradox*, 274.

6. We observe the zero *is* the nine in vortex-based mathematics. In this unique counting system, 0 and 9 are different manifestations of the same number (i.e., every other number is counted/valued by its relationship to 0 and 9). In this way, ratio is fundamental. Can the Nothing be the All without conflating 0 and 9? At the point of merger, the tension becomes the most powerful, the most creative in a sense. As each number (1 through 8) has meaning in relation to 0 and 9, it emphasizes spatiality (or proximity/geometry) because to travel one direction is to go toward 9 and the other direction is toward 0, yet they both ultimately go toward the same place. It is like a torus. A two-dimensional example would be navigating the Marko Rodin circle 1, 2, 4, 8, 7 to complete the first "circle", whereupon at 7 the digit enters the tens (16) so the energy has gone *through* 9/0 and emerged on the other side.

expanding into one extreme (0) you contract into the other (9). If your destination would ultimately be the same either way (i.e., the center point of a torus), then the actions (expansion and contraction) are also ultimately the same. Indeed, sometimes the way to find these math and science examples is by studying extremes, or in the case of vortex-based mathematics an alternate way of counting spirals and decimals.

And here is a fun example: the phi symbol (ϕ) is a combination of the symbols for 0 and 1. Likewise, if you change the shape of the Arabic numeral 0 it becomes 1, and vice-versa. It is quite likely God is having fun with *anything* God chooses.

TWO PARADOXES ABOUT CERTAINTY (EVERYTHING) THAT COMPETE FOR RELATIVE PROBABILITY

A hypothetical space with no causality requires no beginning (in the sense a "beginning" is a perspective on order), and yet without causality that hypothetical space (constituted entirely by Randomness) cannot *experience* itself in the way of observation, so it does not exist to any observer that requires causality to think. Therefore, the first cause is really just the first measurement/observation. Or, the universe is in a perpetual state of fluctuation between Random (noncausality) and Certainty (causality). Meanwhile, free will is an excellent mechanism to produce perpetual imbalance.

A theory presented in this book is that fundamental hypothetical paradoxes are competing for God's attention. The first words of the Bible, "In the beginning" (בראשית) would involve an observing head (ראש) entering (ב) being/existence (ית).[7]

To *create* is to *cause*, so something truly Random is not created, but rather *exists* fundamentally. And if this is true, then something truly Certain is also not created but rather exists fundamentally.

When scientists do tests for randomness, they first eliminate all potential sources of outside *observation*. So, in practical experience, randomness and observation are mutually exclusive.

7. Note that in Hebrew ית is not in Strong's dictionary, but in first century Aramaic it meant "being/existence." Indeed, it is the root of אית in John 1:1, which is utilized in its most fundamental sense to describe God simultaneously having the word and existing as the word.

If an all-encompassing Nothing is the foundation of all concepts and matter, so too would the All and the Nothing be present in some manner in everything that was, is, and will be, including paradoxes.

TWO PARADOXES ABOUT RANDOMNESS (NOTHING) THAT COMPETE FOR RELATIVE PROBABILITY

Absolute and total Randomness is noncausality. Absolute and total Uniformity (certainty) is also a form of the Nothing because there is no way to distinguish or discern anything from anything else. A brain cell requires order and distinction to develop logic. In a world built on perception, *contrast* is the only evidence we have to know anything.

In total Randomness you learn nothing because logical progression is impossible to discern. In total Uniformity, you observe nothing because there is no contrast to stimulate brain activity. This is why we know nothing (or next to nothing) about the fabric of empty space itself—empty space is so uniform that we cannot grasp any logical progression from distinction to distinction. General relativity teaches that no pressure or weight is felt/experienced in a free fall in empty space. Only when there is an acceleration, deceleration, or change in trajectory or angular momentum do we feel pressure or weight.

With Randomness, anything or nothing is possible. And so, raw possibility resides in a state of perceived or hypothetical Nothingness.

How does a thing (or situation) become random? The thing must represent total absence of probability, *or (paradoxically to the extent Randomness and Nothing are the same thing)* total uniformity of probability (the absence of distinction and therefore Nothing, from which anything can randomly emerge conceptually because of the paradox). The Nothing is just one type of uniformity, where the spacetime clock/scale is set to zero. From the human perspective, maybe if we could see behind the zero, we would see a spiral history. What if just as the Fibonacci sequence only works if you start with 0 and 1, so too you unwind that clock and travel back to zero and one as primal concepts that preexist the concept of order itself that is necessary to even create logic and math. In this manner, paradox would likely be more primal than math.

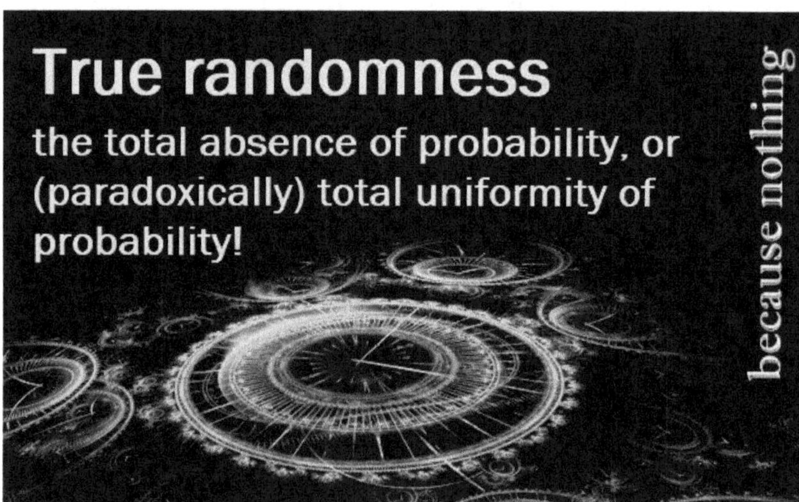

How does a thing become certain? Because of quantum physics, we can now speculate with some confidence this happens by converting probability into observation or interactive experience.

Does Randomness create Order to create Randomness in a self-perpetuating cycle? How does something random come from something ordered? Pi and Phi have seemingly random digits yet represent order. It is speculated there must be clues in them to perpetual motion generation, and it is calculable that artificial intelligence, if unchecked, would unlock exponential insights and new technologies.

PROBABILITY IS AN IF

The power of if/then statements has driven logic throughout generations, and modern computing was founded with the principle. And now in quantum computing, 0 and 1 exist simultaneously in superposition, which makes "if" even more iffy.

Consider the possibility of "if." This is the deepest and most fundamental concept I can imagine. It is the simplest way to identify conceptual/hypothetical Possibility itself. "If" is a question, but in the more fundamental sense of being an inherent hypothetical.

If Hebrew יף (sound "if") means beautiful or attractive. Was the Nothing attracted to "if"? Is it the merger/joining of 0 and 1 that begins the Fibonacci sequence? If 0 then 1. Result: propagation of life.

Because a fundamental paradox can never be fully resolved (whether completed or annihilated), you never reach 1 or 0, so you remain probabilistic. You remain an "if," a hypothetical.

PARADOXES OF ENTANGLEMENT AND CONSERVATION

Probability is an abstraction, and as such it also has a quality of Nothingness.

A paradox is also an abstraction, and paradox is the ultimate form of a question. A question is ultimately a searching. Searching is movement toward the objective of interaction among parts.

The paradox inherently searches in a state of unrest. And it is this idea of unrest that creates the potential for paradoxes to be creative perpetual energy phenomena.

A great question is whether paradox precedes the law of conservation of energy (a paradox never comes to a resting point). If paradox is creative in the most fundamental sense, as hypothesized in this book, then conservation of energy[8] simply describes its character; hence, there should be an opposing force (the other side of the paradox) to conservation of

8. Conservation of energy is a fundamental law in physics, and is philosophically recognized as a principal reason the universe appears just and ordered.

energy. Perhaps it is light travelling every probability through quantum entanglement.

Here is why:

1. quantum entanglement = exponential complexity = expressing 1 (All)
2. conservation of energy = exponential simplicity = expressing 0 (Nothing)

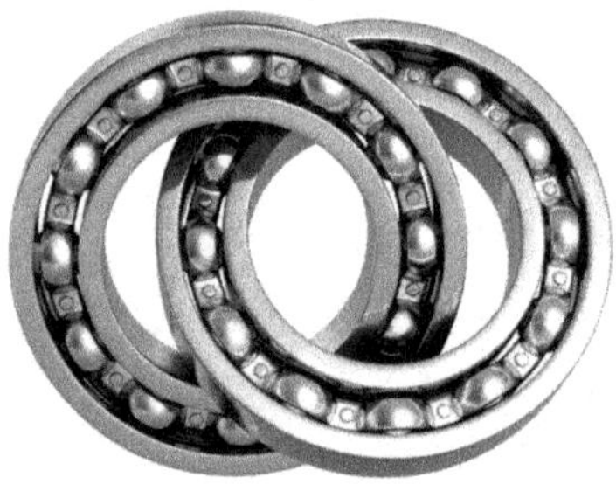

The paradox is that both of these are the same thing!

Entanglement = 1 (probability of 0 → 1)
Conservation = 0 (probability of 1 → 0).

How does God entangle *everything* that is simultaneously striving to be *nothing*?

The paradox is that both of these are the same thing! Entanglement = 1 (probability of 0 to 1); while Conservation = 0 (probability of 1 to 0). How does God entangle *Everything* that is simultaneously striving to be *Nothing*? This is a great mystery, which perhaps the Bible helps answer through commandments (fundamental laws), especially the Greatest Commandment in Mark 12:29–30 regarding oneness, "The first from any of the commandments, 'Hear Israel, the Lord our God is one Lord, and that you love the Lord your God with any/all of your heart and with any/all of your soul and with any/all of your mind and with any/all of your strength.' This is the first commandment." See also Deuteronomy 6:4–5. This is followed by the second commandment that will remind some of quantum entanglement, "And the second [commandment] is like it, that you should love your neighbor as yourself." Mark 12:31.

Reading the Bible, one can find biblical principles that may relate to conservation of energy. For example: emphasizing debt forgiveness and rebirth from a starting point (0) we learn from the Messiah's life to death to life, the renewal of baptism, the resting on sabbath, and the release of jubilee. Also emphasizing conservation of energy are the many biblical passages about accurate weights and measures that ensure equations balance. Avoiding leaven is another example.

The Bible places such a large emphasis on justice, it is observed as a completing force. An essential principle for justice is simplicity first. From Nothing (hypothetical Randomness) there must emerge simplicity first. Otherwise, there would be an infinite multiverse immediately, which is theologically abhorrent as discussed in chapter 4. Simplicity first is the principle of conservation of energy. Notably, the original Aramaic gospel was called Peshitta, which means "simple, straight."

Recall my earlier statement, "the result is that hypotheticals exist because they are easier to allow than they are to exclude." How can we say it is easier to allow possibilities than exclude them if there can be no possibilities without Randomness? "Possible Randomness" is a paradox, and may be source or close to it.

Why would hypothetical Randomness limit itself by a conservation principle? This may be the most important question a man asks because the answer can preempt the possibility of an exponentially evil multiverse. Again, chapter 4 details this moral and legal issue.

The theory of this book is that hypothetical Randomness is the only *primal* force in existence, so only itself can limit itself. This logic calls for God. And indeed, most probably the answer to "why" (see chapter 7) is God's will, because God is good. Christians already come to this answer intuitively, thank God. There is a paradoxical circularity to the great question of why, so it makes sense that intuition would be needed to help resolve the circularity.

The golden ratio is among the best evidence of conservation and simplicity first. 0, 1, 1, 2, 3, 5... Notice the beginning (0, 1) is simplicity first, as it does not get simpler than 0 and 1. The golden ratio is present in all of life and nature. Science has tried to co-opt these to advance suspicious theories of evolution that are filled with hypocritical inconsistencies.[9]

Let us ask, if simplicity first, then why God rather than evolution? As we try to answer this question, it becomes increasingly meaningful and humorous that God has put such a spotlight on evolutionary science in the modern day. It is relevant to our conversations with others (especially agnostics).

If God evolved from limited possibilities (by assumption utter chaos as Randomness is limited possibilities), then justice is secure, because God

9. For a thorough debunking of modern evolutionary theories, see Brown, *In The Beginning*, 1–101, 315–408.

would evolve and reign before any great injustice occurs.[10] In the beginning, in God's existence navigating hypothetical Randomness) perhaps God observed some emerged injustices, and even conquered them. Much of our experience on earth is learning how to subdue injustice.

The subject of law and justice involves political science. In politics, hypocrisies seem to be more common than paradoxes. But there are paradoxes. Being a conservative is paradoxical because not changing would change everything. God is the author of conservation of energy; which is perhaps a sign that God is the ultimate conservative, especially as it is generally believed that primarily conservatives have guarded the Bible over millennia and provided the political and legal solutions to implement its wisdom. The great principle *conservation of energy* is about slow change, efficient change, necessary change. A conservative never reaches the goal to preserve what was, but instead preserves a method of transition. Politics has been no small matter in history; it has determined rights, economies, and wars governing the rise and fall of nations.

EXPLANATION FOR MIRACLES

The illusion universe (sometimes dubbed holographic universe) also explains the existence of miracles, where matter changes via thought alone (see examples with faith healing (good) yet also hypnosis (questionable, often misused form of biohacking)).

As the Bible records with many examples, God is the ultimate miracle maker. This is not surprising as God is the creator of probability, because anything in God is possible. See e.g., Matthew 19:26.

The Messiah, whose title is a wordplay for possibilities, performed many miracles. As we observe miracles, there has always been something unsatisfying to our human minds about a material world that does not

10. The counterargument is the multiverse (which is debunked in chapter 4). Multiverse proponents can theorize a simultaneous hypothetical Randomness where the chaos includes exponential amounts of order just because of the sheer infinite amount of randomness. "Infinite monkeys" is their argument—they say if all possibilities exist in order for God to be present, then the multiverse becomes possible with its exponential injustices. The main problem with infinity/multiverse is utter ridiculousness and injustice: there would be an infinite number of identical universes as well as different universes, and an infinite number of Supreme Gods. The best thing we can say about infinity is it is infinitely illogical and infinitely inefficient. Multiverse theorists attempt to dispense with processes, for if all is infinite then everything exists simultaneously, and therefore change is impossible.

adhere to materialism. Fortunately, our dilemma is resolved by a probabilistic universe where anything in God is possible, including paradoxes.

Why is a paradox not self-contradictory? Because it is made of possibility itself. C.S. Lewis argued that anything self-contradictory is impossible, "I know very well that if it is self-contradictory it is absolutely impossible. The absolutely impossible may also be called the intrinsically impossible because it carries its impossibility within itself, instead of borrowing it from other impossibilities which in their turn depend upon others . . . You may attribute miracles to Him, but not nonsense. This is no limit to His power."[11]

I would have liked to hear Lewis' thoughts on this book, where it is argued that paradoxes are not self-contradictory because at their root they are comprised of the same thing.

HYPOCRISIES V. PARADOXES

Hypocrisy destroys. Much of the Bible is devoted to exposing hypocrisies. The Messiah particularly exposed the teachers of the law, scribes and Pharisees, as hypocrites.

11. Lewis, *Problem of Pain*, 18.

Because Satan is a lying imposter (see e.g., John 8:33, Revelation 12:9), what exactly has Satan been attempting to copy with his hypocrisies? Is Satan trying to copy paradoxes that *create*?

Very little of the Bible is openly revealing that God creates through paradoxes. But if you study dualities, you can begin to learn their prevalence. Ecclesiastes 3:1–8 for example foreshadows how the Messiah's life reveals a paradox—a time to be born and a time to die—in dying he lived. See also John 12:24, "Unless a grain of wheat falls and dies in the earth, it remains alone. But if it dies, it produces much fruit." And the Messiah revealed this paradox from Ecclesiastes—a time for war and a time for peace—he is paradoxically the prince of peace who brings war, resulting in peace. See e.g., Isaiah 9:6, Revelation 19:15.

Satan (including his followers) would try to mock these things by claiming they are hypocrisies. And indeed, the concepts themselves would be hypocrisies when wielded by Satan (i.e., war is peace; death is life).

But the Messiah is of God, and therefore his life proves that even the greatest dichotomies are actually *creative* paradoxes when stewarded by God. Throughout Christian history theologians have openly pondered the paradoxical nature of the biblical account, such as life in death, and divine in human.[12]

A great paradox of law is revealed in the Messiah. He fulfilled the law that causes death to any man who transgresses, and in so doing saved the life of any man who believes in him. There is indeed a great paradox here, mediated by belief. What is the law, if not an abstraction? The law is not a physical hammer, but rather the law is an idea, a concept, an abstract thing. It is quite fitting that belief navigates both the life and death of the law.

US Supreme Court Justice Oliver Wendell Holmes stated famously, "The life of the law has not been logic : it has been experience."[13] Probably the life of God's perfect law is belief, even if Justice Holmes accurately described man's imperfect law.

As hypothesized in this book, paradoxes compete for God's perspective, such that competition is inherent to life. God incentivizes mankind to choose paradoxes (and therefore learn about God) rather than choose hypocrisies that lead to death.

12. See e.g., Kandiah, *Paradoxology*, 238, on Christ as God and man, some say paradox, yet others say "antinomy."

13. Holmes, *Common Law*, 1.

Competition is a fundamental law in the universe. It is also respected in man's law, because we recognize that competition regularly improves outcomes, by incentivizing creativity and productivity. Especially after having observed the consistently catastrophic failures of Communism, many would say that respectful competition is the truest form of the golden rule in Matthew 7:12. Do you want men to give you handouts that dilute your worth and impugn your ability to be diligent pursuant to the many instructions in Psalms and Proverbs? Or do you want men to respectfully compete in economies that incentivize people to provide for one another and to be successful providers? Even in this we find a paradox about the community benefit of being honorably selfish (diligently caring for self and family).

TEST YOUR SKILLS: PARADOX OR HYPOCRISY

Paradox: two true statements that appear to conflict, yet are creative and ultimately the same thing.

Hypocrisy: portraying one thing but then contradicting yourself, to gain some undeserved advantage from misperception.

Examples of Paradox:

- If an unstoppable force meets an immovable object, it is because they are the same everything, the same nothing, and/or perfect opposites. Go ahead, try and stop "everything." Try and stop "nothing." Try to move "everything" (move it into what and where?) or try to move "nothing."
- Zero and one are the same number. There is only one zero (paradoxically, if there is no zero, then there is zero). Oneness experiences zero contrast, as nothing exists apart from Oneness. Therefore, Oneness is invisible or purely conceptual just like zero. When Oneness is perfectly uniform, it learns nothing about itself and therefore only exists and experiences in relation to zero.
- An agreement to agree is illusory.

Examples of Hypocrisy:

- A priest mistranslates the Bible to claim he is eternal.
- A salesman sells a perpetual motion engine without a lifetime guarantee.
- Politicians agree on inalienable rights endowed by the Creator, and then claim the power to disregard those rights to advance lesser laws and policies.

There are real world effects from paradoxes and hypocrisies. These are not purely esoteric matters.

Chapter 3

First Cause Paradox

SUMMARY

THROUGHOUT RECORDED HISTORY, MANKIND has asked "What was the first cause?" The question can mislead because causality is not primal in the universe. Cause and effect are a finished product (or result) composed of these much more fundamental concepts predating causality: God's will and fundamental laws that create order in time and space. And in these more fundamental concepts, *faith* is paradoxically real. So, the better question is, "What is the first or primal faith?" The answer is probably God's faith in himself, such that God literally is faith that co-emerges in hypothetical Randomness. Hypothetically, God's faith would create God's self as order (or discernable ratio) paradoxically from inside conceptual Randomness.

Relevance for your life . . .	Imagine for your life . . .
Your faith increases as you understand God's will is *the* reason you are alive. Cause and effect in the material world help describe you but do not create you. This is why God, rather than the universe, is your Father. God is teaching you on earth about the primacy of will.	Your life improves when you put things in the right order, like words in a sentence, notes in a song, and you in your family.

Logical Hypothesis

1. If Randomness,
2. Then no causation,
3. Therefore no first cause.

FIRST CAUSE

This book invites readers to imagine we exist as hypotheticals, meaningful illusions, in the mind of the One, who is simultaneously the Nothing. This paradox between All and Nothing would create the illusion energy we call reality. I submit this inherent paradox is superior to the red herring question, "what is the first cause?" Philosophers throughout the ages perhaps misled us with that question by calling it primal and unanswerable. Once we see through the definitions of time, we can plainly observe the "first cause" is not primal. Time depends entirely on order, which is dependent upon God's law, and is the product of God's will.

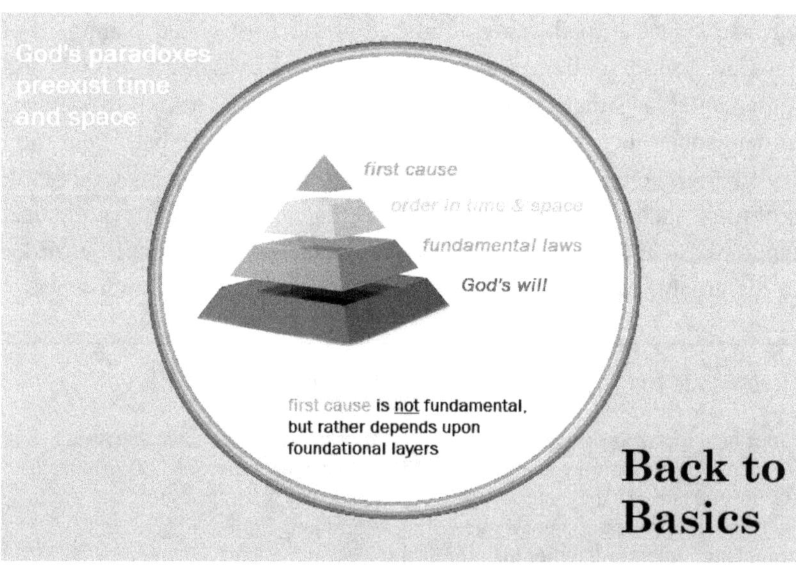

In other words, the question of the first cause is not primal because "cause" is just a description of *order* in time and space. And order is *law*, such as universal laws. And law is an expression of *will*, God's will of course. And that will can be expressed in many ways (i.e., frequentist is the most common method to human perception; probabilistic is more primal as we have learned in the quantum physics era, and allowing of miracles/improbabilities).

Therefore, you can hypothesize the first cause is better described as the paradoxical relationship inherent to the Nothing and the All, which is navigated by faith. And even still, faith is a description of something more primal: Oneness. Kingdom is a description of Oneness. It is highly paradoxical that Nothing is a fundamental form of Oneness, because Nothing is complete in itself.

What if God (the ultimate kingdom) divides himself within his own Oneness such that we perceive order among the parts/perspectives? God is often analogized to light, which is divisible as the Bible states in Job 38:24. Incidentally, the Bible correctly predicted this science thousands of years before Isaac Newton proved it (by separating white light then recombining it). The division/sharing of God in this thought exercise would be probability (as we remember the word Messiah (משיחא) in the language spoken by Yahshua is a wordplay for "probable/possible" (משכחנא)),[1] because the nature of God is unity (oneness, kingdom). So, the result is that anything in God is possible and God uses probability to reveal the world (all light is probabilistic). If we focus only on the face of things, we miss the deeper probability in all.

In Aramaic the word "hypocrite" is an expression derived from the literal words "taker of face." Probably the word origin was someone who only cares for the face of things, implying the face is different than the substance. As we live and judge as hypocrites (takers of face), is it any surprise we most clearly see God alive and judging on the face/edge of an age (i.e., Exodus, Revelation)?

There is freedom in the expression anything in God is possible. As long as there is freedom, there will always be a need for faith. Thus, there is freedom in faith, and faith in freedom. They are paradoxically the same. Many of these paradoxes are intuitive to Christians but are entirely foreign to non-believers.

Once you have the key to unlock paradoxes, you can point to any two opposites and describe how they are actually the same thing at their core or extreme. What is this "key"? Is it necessity? For example, light needs dark in

1. Smith, *Syriac Dictionary*, 306 (משכחנא possible and משכחנותא possibility).

order to define itself as light, for there to be the contrast necessary for observation. Without contrast there is no order, and all is random meaning all is nothing. But necessity is not the whole key, just a part of it (and a helpful one to human logic seeking truth). For necessity is determined by something deeper still—will. God's will sets the rules for whether something is necessary or not, related or unrelated. Necessity itself is a description of entanglement or dependency.

UNANSWERED, UNQUESTIONED

Is the concept of faith self-emergent from the exclusive Oneness of the Nothing? In the previous chapter, I attempted to answer these questions with a logically satisfying response:

1. What are two paradoxes about randomness (Nothing) that could theoretically create a charge between them to compete for relative probability?
2. And what are two paradoxes about certainty (Everything) that could theoretically create a void as they compete for relative probability?

The context for these questions could be that in quantum mechanics, light and matter are described probabilistically. It is possible their balance can be defined by two paradoxes competing for relative probability (power) as a single paradox: (A) the most *unlikely* idea to our minds is the absence of a first cause, and (B) the most *likely* idea to our minds is the presence of self (i.e., our experience in this moment is really happening).

Neither idea (A) or (B) can be proven with absolute certainty, so each remains in a state of motion toward its extreme. Therefore, perhaps we experience a duality of paradox through probability mechanics. As the famous physicist Niels Bohr stated, "There is no quantum world. There is only an abstract quantum physical description."[2]

Things with the same answer suggest that somewhere beyond the surface you can find the same question.

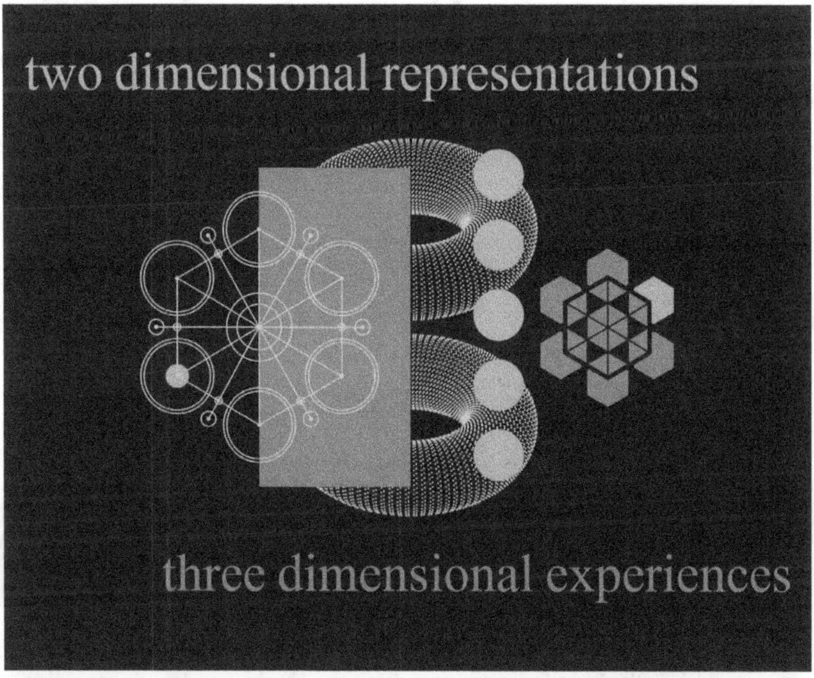

THE MOST POWERFUL FORCE IS THE NETWORK

Surely it is no accident the known universe looks like a brain (neural network).[3] It appears to be a great sign or clue from God at this time in scientific knowledge. Because the observable universe is shaped like neurons,

2. Petersen, "Philosophy of Niels Bohr," 12.

3. See Pappas, "New Dark Matter Map," para. 1. See also the gravitational simulation model. Springel, "Largest N-Body Simulation," figure 1.

it evidences intelligence in *everything*, which is why everything is meaningful and connected. Intelligence is connection, pattern.

Boni recognizes this from quantum physics:

> "[T]he fabric of physical reality is *not physical at all*. In fact, that which serves fundamentally, foundationally, and elementally as the basis of physical reality is more akin to 'abstract cogitation' than anything else . . . Quantum physics reveals that physical reality literally reduces to a mentalistic essence not unlike that of 'thought', or 'conception'. Beneath all physicality there exists a foundational quantum substrate that is wholly incorporeal and akin to the conceptual slate of cognition whereupon the rational mind does its bidding. As Sir Arthur Eddington simply but straightforwardly put it, 'The stuff of the universe is mind stuff.' [and physicist Sir James Jeans] 'The universe begins to look more like a great thought than a great machine.'"[4]

From a materialist viewpoint, the network is arguably the most powerful known force in the universe. For example, a network of brain cells (brain) is more powerful than a single cell (neuron). A network of circuits (computer) is more powerful than a single circuit (gate). A network of people (kingdom) is more powerful than a single person (citizen). As humans, we have learned this lesson the nice way and the hard way.

The modern world pushes men to reduce themselves to two-dimensional objects (surfaces like flat screens and texts), and creates elaborate networks governing two-dimensional objects (networks like courts that execute written laws, and media that executes scripts) to lord over men. By contrast, God invites man to join the assembly of Christians, so God may enter and give purpose to everything within. This elevates the three-dimensional (humans) to multidimensional (humans + God).[5]

As with solitaire, weak men gravitate to simple puzzle solving, allowing themselves to *become* the two-dimensional puzzle, and therefore lose the game by winning.

By contrast, a successful human grows upward through dimensions to realize the full potential of paradoxical living in God, where the human

4. Boni, *Quantum Christian Realism*, 4–5, 71.

5. Just as cartographers cannot represent a three-dimensional globe accurately on a two-dimensional surface (i.e., if direction is consistent then shape and areas are distorted), so too with trying to represent multi-dimensional possibilities (waves) in the materialism of our minds (i.e., Heisenberg uncertainty—if speed is accurately represented, then location is distorted).

dies (on earth) to gain life (in heaven). There are some paradoxical aspects of rebirth, such as: you cannot choose to be born or choose to be saved (per John 6:44), but you can choose to be faithful and receive life (per Deuteronomy 30:19).

C.S. Lewis wrote, "The size and emptiness of the universe which frightened us at the outset of this book, should awe us still, for though they may be no more than a subjective by-product of our three-dimensional imagining, yet they symbolise great truth."[6]

WHO IS GOD? DESCRIBING OUR RELATIONSHIP WITH GOD TEACHES US PARADOXICAL ANSWERS.

Theoretically, human faith, logic, and skepticism can all be manipulated beyond human comprehension. So, why should we trust our minds? This is a kind of nihilistic question, and it often feels ignorant of experience. By contrast, it feels honest to say that people observe the power to think and tinker, and because humans did not create this world, the Authority that did is showing us something.

The Limits of Skepticism Teach Us Faith Is More Fundamental

Our minds are fragile due to limited perspectives. For example, no human can prove whether ancient invisible powers (spirits, aliens, artificial intelligence) may be fooling us about existence. Indeed, the idea that we live in a computer simulation has been popularized for laymen (i.e., the movie The Matrix) and in academia (i.e., author Nick Bostrom).[7]

Consider hypothetically what trillion-year-old aliens or artificial intelligence could orchestrate through mind control that utilizes hidden physics. As we have limited perspectives, we must admit logically that the extent of the mind control is theoretically unlimited—time and space might be just the tip of some exponential iceberg of deception. When faced

6. Lewis, *Problem of Pain*, 159.

7. University of Oxford professor of philosophy Nick Bostrom argued statistically we are more likely living in a computer simulation. A critique of his theory is beyond the scope of this book, but I think ultimately he followed the wrong trail to a dead end of materialist philosophy, together with a secular assumption bias. I would argue the correct trail, as stated in this book, is the probability that we are living inside God's head, between hypothetical paradoxes.

with any classic "alien powers" example, we are required to fix our minds in one kind of thing—uncertainty, doubt, speculation, human incapacity to answer—skepticism. So, in this example, we find yet another proof that we cannot define capital "G" God, nor can we define Everything or Truth.

All we have are limited powers of logic, faith, and skepticism. And we exercise these powers in a world that reduces to wave functions.[8] And logically, the more fantastic the world and heavens appear, the more likely our human minds will speculate that our earth-based chemical neurons are the subject of unnatural interference—this is because we earthlings observe every day that our natural world of chemicals is not fantastic, but predictable. Factoids that tend to disrupt our logic these days come from things like quantum mechanics (which most humans do not study), as previously "impossible" and bizarre observations appear commonplace.

All of light and matter are described probabilistically. In layman's lectures on QED, physicist Richard Feynman showed how the path of a particle of light represents all possible pathways originally available to that particle geometrically, so the ultimate pathway travelled is predicted by a probability amplitude of what is possible. As Feynman explains:[9]

> "the probability that a particular event occurs is the square of a final arrow that is found by drawing an arrow for each way the event could happen, and then combining ('adding') the arrows . . . A photon that reaches the detector has a nearly equal chance of going on *any* path, so all the little arrows have nearly the same length." Pages 39 and 41.
>
> "Does this mean that physics, a science of great exactitude, has been reduced to calculating only the *probability* of an event, and not predicting exactly what will happen? Yes. That's a retreat, but that's the way it is: Nature permits us to calculate only probabilities. Yet science has not collapsed . . . Underneath so many of the phenomena we see every day are only three basic actions: one is described by the simple coupling number, *j*; the other two by functions—P(A to B) and E(A to B)—both of which are closely

8. As observed by physicist John Wheeler, "No elementary phenomenon is a real phenomenon until it is an observed phenomenon." Boni, *Quantum Christian Realism*, 71. Now recall the famous expression "you are my witnesses" in Isaiah 43:10, "You are my witnesses says Yahweh, and my worker whom I have chosen, that you may know and believe me, and observe that I am he. Before me no god was formed, nor after me shall there be."

9. Feynman, *QED*, 19, 39, 41, 120.

related. That's all there is to it, and from it all the rest of the laws of physics come." Pages 19 and 120.

This may or (more likely) may not be the end of the matter. The point is that all scientific endeavors are meant to refine our analysis of likely probabilities that a particular phenomenon is, or is not, a reality for our minds. From there, we have logic, faith, and skepticism—the human experience.

Faced with this imperfect dilemma of probabilities and logic, should we therefore strive to remain logically and emotionally impartial to everything we observe (like laboratory subjects), so that if humans are ever confronted with something perfect, we will be more likely to know, and gravitate, and yield to the perfection observed? No, that is just another question we cannot answer. And the strategy of denying our own logic and emotions defies common experience that our bodies and relationships thrive when we listen to ourselves (as explained deeper in chapter 7).

When a man cannot logically determine an ultimate answer, he must decide whether he will endeavor to apply another tool capable of seeking his desired ultimate answer. So, where man's logic falls short, many emotionally-exhausted humans embark on the life-long process of conditioning their minds around the conclusion that faith/trust can prevail, because faith's circularity is peacefully complete, whereas the process of logic seeks to rely on either (1) ever-changing evidence in the hope of building reliable conclusions upon unexplainable phenomena (which is an illogical and/or incomplete process), or (2) mathematical and/or intuitive probabilities that help us feel comfortable rationalizing our world views (which is logical toward limited pursuits, but probably illogical toward any goal of finding ultimate answers to questions like "what is truth").

My own personal conclusion is that there is a logical way to approach faith and nihilism, and an illogical way to approach these concepts as well. First though I should say something practical and down-to-earth: my common sense advises me to avoid crime on this planet and be fair and humble, and try to practice love appropriately.

When a man sincerely believes that faith is accomplished through love, and that God is love, he will most likely try to give his faith freely to his mind's appreciation of God. In this way, a psychiatrist could say he falls in love with his own mind. And perhaps there is some great truth there; we do not know. In any case, based on human experience a faithful man is likely to accept his preferred kind of signs in the physical world (i.e., synchronicities, measurements, saints), as well as messengers in the form of other humans making prophecies (much of the Bible), and spirits (unexplained energies).

I believe God speaks in signs and synchronicities to encourage us to spend more time thinking about his messages, and realizing the possibilities inherent to each message. Our written words are good but they are just like shadows of a greater living language, as evidenced by the aleph bet flame letters (explained by researchers Tenen and Winter) referenced later in this book.

What if logic, faith, and skepticism are all the same thing, because everything is faith? Can skepticism be just another form of faith? People have always observed this about atheists, that they are *religious* about their assumption-filled skepticism. Logic is a function of probability occupying the middle space between faith (belief) and faith (skepticism).

How do men draw the distinction between belief and fear of nonbelief? From one perspective, a man can perfect his faith in ignorance, but not logic, unless by quantum entanglement he paradoxically enjoys both.

Another way to analyze religion is whether mankind will find itself called to answer a more specific and narrow question—who can save mankind? And given all the evil in the world, one might logically think the ultimate Authority would indeed come to save us. You might even say that the worse things get here on planet earth, the more likely we are to believe that whoever saves us is someone important—so, the greater the danger, the greater the Savior. And an analogous phenomenon—the greater the forgiveness, the greater the love. See Luke 7:47.

In the next chapter we will explore the most frightening question of all—whether the ultimate Authority of the universe would question whether he too was a created being. And if so, would this Authority theoretically manufacture injustice in the universe just to see if there were a *greater* Authority willing to intercede? If there is a more frightening question in the known universe, I have not pondered it.

Now for a thought exercise: imagine you are God with the keys to undo all of creation—not only all that will be, but all that ever was. You second-guess yourself before pulling the trigger. You question whether the injustices in the universe are outweighed by the worth of the moments of joy that can only exist through tenuousness. You doubt your own power to destroy everything (including yourself), because once you do, you have no way to verify the job is complete. Indeed, you would need faith in yourself to do the job. Almost hoping you lack the power to erase everything, you rationalize that if everything over which you had the power to erase was only a fiction (or illusion), there would be no injustice. Indeed, if you did erase all the fiction in the world, perhaps all that would be left would be reality, such that you would only succeed in erasing the only escape from injustice. Feeling overwhelmed, perhaps you decide to play some music. Is this what God is currently doing? How on earth could we ever know?

The Limits of Logic Teach Us The Meaning of Choice

Many readers of the previous section may comment, "So what?" Perhaps you have experienced many of these same nihilistic thoughts yourself, and you know them to lead nowhere. With that said, allow me to add that even though humans may lack faith at times, we do seem to have a decent grasp of our intuitive ability to recognize burdens and benefits in both physical and conceptual matters, and so we stand at least a chance to gravitate toward a benevolent provider and feel good about our choice, or at least feel good about the risk we took even if it turns out we later feel we made the wrong choice.

It is also thought-provoking that an all-providing heavenly parent would logically be the most likely candidate to receive faith and love from humans. The archetype fills fundamental roles: creator, resource provider, educator, protector, enforcer of justice, and offeror of parental love. As humans, many of us are fortunate enough to know the experience of having caring parents as well as being a caring parent. This experience, along with a sense of the blessing of free will, can help open our hearts to the possibility of a "parent" coming along, giving us milk and cookies, and telling us that we are their child, and we are loved. Is holding on to human skepticism worth rejecting the opportunity to have a relationship like that, even if it is ultimately with some kind of alien or artificial intelligence in the universe? This is what a skeptic would ask, and to the extent we have free will, perhaps we are making the choice in every moment.

Humans gravitate toward resources. If a living thing, or even a corporation, gives us valuable resources we tend to respect and honor it. And when those resources are withdrawn, we also withdraw. This archetype is pervasive, for example it easily dates back to the ancient agrarian societies who worshipped fertility gods. And it continues today, as many choose to give their allegiance to whomever gives them resources, such as the government, supermarket, family, charity, and so on.

We often see destruction as the opposite of resources, even though the earth shows us sometimes the best way to replenish fertility is with a flood, volcano eruption, wildfire, etc. In the Book of Revelation, it is said that many will curse the one who brings destruction and punishment upon the earth, and these men will not repent. Revelation 16:9–11. Things are not always what they seem.

Because we are all so different on earth, it makes sense that some of us want puzzles and codes to decipher, whereas some of us will try to demand immediately clear and unflinching evidence of all truth. Some of us want adventure challenges on a rugged planet, and some just want to be transported to a heavenly paradise right now. If you are a word nerd like me, you smile at the wordplays between paradox and pair of ducks, paradise and pair of dice. Our Creator would naturally know our character and would design a system for us to bring out the best of our best in many categories, as this would bring the Creator the most honor. Do you see the mutually-beneficial relationship here? This is the psychological key! Trusting in something mutually beneficial for the Creator and the creation is a logical place for one's hopes and joy. It is the most logical way to make a choice over which you will feel positive, even if it turns out later that you made the wrong choice. This is game theory, and throughout the ages it points to faith in God as the correct answer.

Paradoxical Proof of Faith

You need faith before you have faith. A good verse explaining this paradox is Mark 9:24 ("I believe my lord; help my lack of faith.")

The paradox teaches us the importance of primacy, of order. Recall this chapter title, First Cause Paradox. Cause itself is not primal. The more primal question is "What is the first or primal faith?" Probably the answer is God's faith in himself. This book hypothesizes God is faith that co-emerges in hypothetical Randomness. In this hypothetical, God's faith would create God's self as order (or discernable ratio) paradoxically from inside conceptual Randomness.

In any case (or back down to earth), we do experience faith is paradoxically real. And it is something we need. So how do we acquire this self-fulfilling quality? The Bible is the primary method to inform our faith. It works like a proof of faith.

יהוה is the Most High and fundamental to universal creation by way of an ability to control physical and conceptual phenomena like time and math and quantum mechanics. The ancient and literal Bible describes that God created a powerful race of advanced beings in the universe who created earth and go by the name אלהים (which in modern Hebrew is "Elohim"). In this view, the Elohim kingdom would be what is popularly referred to as the kingdom of heaven. But the Bible also describes something called "the heavens of the heavens" (1 Kings 8:27), so it is complicated.

The relationship between יהוה and אלהים may be ambiguous from the literal Bible alone, and perhaps this helps explain verses like Proverbs 25:2, "The honor of Elohim is to hide a word; and the honor of sovereigns to search a word."[10] Here is a paradox: if everyone is sovereign, then no one is sovereign. In the meantime, mankind tends to elevate principles (laws) above each sovereign. This makes words more sovereign than man, and perhaps there is a great lesson there. "In the beginning was the word . . ."

It appears the ways and means of the highest Elohim are staggeringly supernatural and conceptual. If earth learning and survival is our example, then we should approach the Elohim carefully, respectfully, and with an attitude of compassion. I think it would probably amount to undesirable pandering if we just bowed down to someone we could not identify and did

10. The Hebrew word origin for "search" refers to curiosity, one of our most important attributes.

not understand. The example set by the Messiah is essential here. To know God, he lived in harmony with God.

The Elohim with authority over all of us is יהוה, the one who rescued the Hebrews from Egyptian slavery as part of the great saga unfolding on this planet. In the Bible, God repeatedly reminds the Hebrews of God's identity in this regard, the deliverer. It appears the Messiah came to earth on a specific mission (narrow path) from his Father, rather than a mission to explain everything to everyone. John 16:12; Matthew 15:24. The net result is that we are exercising our minds, and building character.

What is Satan's adversarial role in this exercise program? And what was his role in creation? We read Satan has chosen an unrepentant, murderous, and vice-indulgent lifestyle, and experiences God's punishment. In the book of Job (describing events thousands of years ago presumably), Satan was still in good enough standing to enjoy council with God in heaven, but it is said in the Book of Revelation that he will be expelled from heaven after warring with the archangel Michael, and some research suggests this has already occurred and Satan has *some* power today on earth.

The gospel gives this analogy—Satan plants seeds of sin, and the Father plants seeds of virtue. The Messiah ישוע (whose name means Yah's rock (Aramaic) or Yah's fortune (Hebrew)) is the Messiah through whom יהוה provides for the forgiveness of any sins (but not all sin), and יהוה will do justice on earth.

As Christians, we recognize that the kingdom of heaven is powerful and that God is worthy of our trust—that is the faith side of the equation, but I do not consider this a call for blind faith, because we are provided resources and light on earth to see and test evidence. And we engage in risk analysis. I consider it likely (and a well-calculated risk to deduct) that one of the goals of humans is to learn from the Elohim, and if you can, strive to gain entrance into their unique culture/kingdom. So, this planet is analogous to a classroom[11] or playing field with useful knowledge, resources, protection, and lots of freedom (which evidences our creator loves us). And your behavior is measurable, such that whatever you measure will be measured back to you. See Matthew chapter 7, including the golden rule.

The Messiah did not follow all the Bible interpretations of his day, nor did he distribute Torah manuscripts and say "these are perfect." It was the Pharisees who acted like their ink was perfect. And as far as I can tell, the

11. The Bible refers to the Messiah often as teacher, and quite literally the word for disciple means student.

Messiah did not teach much about hierarchies among the Elohim. Rather he healed the sick, helped the helpless, exposed hypocritical religious teaching of Torah-thumping priests (Pharisees), and forgave people. "Fundamentals first," you might say. Be a student of that example and see where it leads you—probably a narrower path than carnal freedom, but ultimately more rewarding if you have some faith/trust you can bring to the covenant.

Church history reveals many wayward religious authorities, upon whom Yahweh executes punishment. See e.g., Ezekiel 20:25–26 ("And therefore, I gave to them statutes that were not good, and justice by which they could not live. And I pronounced them unclean because of their gifts, in offering up any of their firstborn, that I might make them desolate, and they know that I am Yahweh."); Jeremiah 5:31 ("the prophets prophesy falsely and the priests rule by their hands"); Jeremiah 8:8 ("false pen of the scribes"); Jeremiah 14:14 ("The prophets prophesy lies in my name"); 1 Kings 22:23 ("Yahweh has set a lying spirit in the mouth of any of these, your prophets. And Yahweh has declared evil against you."); Isaiah 10:1 ("Woe to those who decree wicked decrees and who write trouble they have written"); Isaiah 28:7 ("priest and prophet wander in strong drink"); Ezekiel 14:9 ("And if the prophet is deceived and speaks a word, I Yahweh have deceived that prophet, and I spread my hand upon him, and destroy him from among my people Israel."); Hosea 9:7 ("the prophet is a fool: the spiritual man is insane, because of your great iniquity and great hostility"); Jonah 3:4–10 (false 40 day prophecy); Micah 3:5–11 (prophesying for money); Matthew 16:22 (even after following him, Peter continued to falsely believe the Messiah would never die); Matthew 22:29 and John 5:42 (on the erroneous teachings of the Pharisees); Matthew 7:15–20 ("beware of false prophets . . . by their fruits you know them.")

I think we are called to discern verses, lest we believe our duty is to continue stoning people, and other violent acts that the Messiah did not do.

Bible translators routinely customize scripture translations to match their theology. It is a phenomenon (sometimes a business) that has spawned much division in Christianity and Catholicism. Most people (children included) know little to nothing of the history of mistranslations, edits, book selection and destruction, papal dictates, etc.

Shall we ignore these historical experiences? The Messiah instructed, as witnessed in the gospel, that he does not want elitism for us here on earth (Matthew 23:8–12), but rather, he wants service (Matthew chapters

5–7). A bad Bible translation serves the same master today as it did in the Messiah's time.

In the next chapter we will expose the worst mistranslation of all: "everlasting hell." Fortunately, the ancient text in Hebrew, Aramaic, and Greek all agree that although hell is real, it is *not* everlasting. Hell is a place of punishment and utter annihilation. God is just and all his punishments are proportional and finite. Frankly, one should not need a biblical scholar to explain how wrong it is to tell children that they will burn for infinity if they do not follow a Bible translation. Yet that is exactly the power of false translations today; they cause people to believe God is a monster who torments people for infinity. Infinite torture is the height of insanity, just as the Old Testament explained repeatedly about various wayward priests and scribes who taught lies to the people. See e.g., Hosea 9:7 ("the prophet is a fool: the spiritual man is insane, because of your great iniquity and great hostility").

An example of the need for reproof: the apostle Peter had been convinced from his childhood that the Messiah would never die (Matthew 16:22), and had to revise his theology when the Messiah corrected him on it shortly before the crucifixion. Christians should always be ready and willing to do an about-face regarding the false doctrines of men whenever and wherever we are confronted with truth. Ezekiel 33:14–16.

Although God is patient and instructive like a parent or teacher, if you continually and unrepentantly violate standards he gives you personally (break key covenants), God will kill you or punish you as your covenant requires, because God loves justice more than he loves limitless clemency for his children or creation. As highlighted, this is a good thing because it evidences consistent adherence to cause and effect. The universe would be utterly terrifying if God did *not* execute justice.

The literal Bible and earth history serve as evidence that no other being has trumped the will of יהוה, and that our God consistently fulfills his obligations under the covenants he makes with mankind. The most that can be said is that sometimes יהוה appears to be persuaded to change an outcome. Jonah 3:4–10.

I am not sure what to make of the destruction of Canaan for example, but if יהוה ever failed to fulfill one of his own covenants, some have theorized there might be a higher power or being above him who would intercede. Based on historical experiences and records of mankind, it seems to be deadly unwise to bet that יהוה will not live up to a covenant with mankind, or that there is a higher authority.

It would feel like pandering to not admit skeptical thoughts. I do not want to be a man of little faith (Matthew 16:8), but I support the idea of being like a humble child trying to discover truths without being deceived. I think a key message of the gospel is that because we are like children, we need to trust God to lead us to truth, and the Messiah is inviting us to follow his example on the narrow path.

Analyzing Abraham's Faith

Mainstream Jews, Muslims, and Christians tend to assume that Abraham was morally righteous for his faith, and that Abraham believed it was proper to offer his son Isaac as a blood sacrifice based on the advice of a messenger.

First, the literal Bible does not specify *who* this messenger was other than to say the messenger was from אלהים ("Elohim"). Many angelic beings (upright and fallen) are described in the Bible as being Elohim (part of the heavenly kingdom with many angels participating in the administration). Even Satan whom יהוה created, has been described as Elohim. See e.g., Job 1, and then also references in Ezekiel 28:14–16, Isaiah 45:7, and Genesis 6.

Personally, I do not think it likely that Yahweh was the Elohim who spoke with Abraham in this encounter and told him to sacrifice Isaac. Indeed, I am reminded of passages like Deuteronomy 12:30–31 and Jeremiah 19:4–6 where Yahweh is said to abhor certain sacrifices. So, I would consider it more likely that Abraham was tempted by one of the darker angels in the kingdom of Elohim. My understanding is that יהוה is not easily described as just one of the Elohim as there are much more specific and reverent ways to identify him. These ways are not always used, but they are used enough to help us test assumptions.

Second, I think that Abraham's choice was morally wrong, but it may be that the two-way idea of right and wrong are not as fundamental here as the three-way idea of choice combined with action and consequence. I mean to say that from a theological standpoint, it seems more likely to me that Abraham's actions were not about right and wrong in that one instant of time, but rather about how much Abraham and his descendants could learn about the results of his choice. In this view, we can assume or not assume that Abraham had free will. Assume he made the choice to pursue a human experience that involved sacrificing another human. If so, then naturally he had other choices, and those choices at least theoretically could have unfolded an entirely different human history.

In any case (regardless of free will or fate), Abraham's choice/action of attempted sacrifice then received the consequence/reaction of a human history where Abraham's descendants constantly grappled with sacrifices (human and animal) and found both tribulation and meaning in their wake. For better or worse my common sense tells me Abraham's choice was not righteous, let alone inherently good. It was only a choice he made, and his decision had consequences in his lifetime and beyond to the present day. Yahshua the Messiah was a descendant of Abraham in one sense, and his activities (most notably submitting to sacrifice on the cross) would bear witness to Abraham's choice. Is it destructive hypocrisy or creative duality that Yahweh allowed Satan/Jews/Romans to kill the Messiah, but did not allow Abraham to kill Isaac?

Third, the religious institution of the Pharisees that Yahshua criticized was very invested in sacrifices. That institution grew on its own traditions, and even spawned businesses revolving around breeding, caging, and harvesting animals for sacrifices. That institution of sacrifice welcomed the consequence of the decision made by Abraham. In this manner, can we say that bad fruit came from Abraham's decision? I think so, yes. And I shudder to think of the sacrifices that go forth from underground Satanist institutions today that borrow heavily from biblical and occult lore.

Fourth, I do not think the translated Bible and Catholic church canonical decisions are perfect.[12] I think dark hearts do evil deeds, and darkness has infiltrated Bible history. In my own imperfect quest to seek out accurate historical accounts, I tend to think it more likely that Yahweh saved Abraham's heinous act by first stopping it (Genesis 22:12), and then using it as a lesson (opportunity for reform = good fruit). Abraham's faith might also be seen in a typological sense of ordaining additional meaning/symbolism in the biblical Messiah's existence and crucifixion.

I think these are mysterious matters, and any man who claims to offer hard conclusions or rituals in this area may be deluding himself, not to mention that he could very well be dangerous if he continues to shed blood sacrifices. In Matthew 9:13, the Messiah specifically says he does not want sacrifice, and to go learn what that meant. He also led by example by setting free the animals in the temple that were in cages awaiting sacrifice by priests. John 2:15–16.

12. And for an interesting perspective on the Lord's character, see Magno, *Suffering is Not God's Will*, 1–14.

I would like to feel comforted (physically and spiritually) on this planet filled with creatures vying for resources, and I would like to know that there is meaning and justice built into our very existence and rhythms. Ritually slitting the throat of another creature does not appeal to me, but not because the action itself and the consequences of such an action have very extreme physical and psychological consequences. I do not feel that death or life cycles are inherently evil or that extreme behavior by itself is inherently evil. Rather, I simply think these things have large consequences and therefore invite heightened scrutiny by watchful eyes and logic, ready to do the best justice we can. And in the case of ritualistic sacrifice, I feel it does indeed invite heightened scrutiny regarding the meaning of justice, and application for justice. So that is what I have attempted to do here with my words—scrutinize Abraham's actions for the purpose of both self-reflection and analysis of societal judgments and conscience.

Seeing the world today, I wonder if Abraham would make the same choice again with young Isaac. In any case, it may be that his choice remains his own separate experience until I take it upon myself to decide upon its value or morality, as I have done here for better or worse. I do not condemn Abraham given the circumstances of an unknown being interfacing with him. All the same, sacrificing another human sure looks like a sin to my logic, and if I were observing Abraham, I like to think I would have stopped him (as God did in Genesis 22:12). It is written in the Gospel of Matthew that we should forgive sins if we wish to have our own sins forgiven. Indeed, even as anger in the heart is its own kind of slaughter, I certainly know that I will need forgiveness.

And yes, this topic of choice, action, and consequence is also supernatural as we consider the origin of the universe and the ability of any life form of consciousness to create an inherent purpose and enforce justice. As empowering as words can be, they are also naturally limiting in our endeavors to discover origin and meaning at universal levels of extreme physics and conceptual logic. Scientifically, we see we can reach out with our tools and describe perceptible materials and concepts for ourselves—almost capture them for ourselves to deem ourselves successful in our thinking. Outside of that, there is the logical possibility that some things may be wholly or partially outside the time and space dimensions that form/limit our very logic, and which are not accessible to scientific brains trying to capture them.

This experiential observer problem is that the very act of using a tool to observe a phenomenon may prevent its observation, or may delude oneself out of the opportunity to receive something that simply is—a realm that only faith or total submission can touch. If it was in this realm that Abraham was "righteous" or seeking righteousness, perhaps a sacrifice with a knife (tool) would do no good whatsoever. In such a hypothetical, would it be relevant then that Abraham was ultimately asked to put down the knife? In that hypothetical, the logical answer is yes.

Paradox or Hypocrisy of Persecution

Consider Proverbs 25:21–22, "If your enemy is hungry, give him food to eat. And if he is thirsty, give him water to drink, so you take coals upon his head and Yahweh rewards you." And Matthew 5:11, "Blessed are you when they curse you and persecute you and they say about you any evil word in falsity because of me. Rejoice then and be glad because your reward is great in heaven." Together, these verses can be misread as perverse, where you are actually helping someone by victimizing them (because then they have reward in heaven). In such a system, if you really love someone, then why not become their enemy and harm them so they get rewards in heaven? And if you do that, then I suppose you hate yourself? Where is the virtue in creating a future kingdom filled with victims of crimes? That sounds sad, not beautiful and glorious. Bible critics would further ask if mankind was created with emotional rage and put on a survival-of-the-fittest planet so we could all be gladiators in some bizarre two-level death match: where people try to serve one another but also empower their enemies (possibly supports injustice) on level one just to see who can accumulate the most treasure in the afterlife on level two.

I suspect there must be a paradox here to resolve the question. To find the paradox, we realize that persecution is a great teacher, directing us how to heal. Suffering is sometimes a gift. It is the leper who cannot feel his limbs who suffers the most, because he cannot learn what is needed to change and heal. This subject is covered in more detail in chapter 7 ("Paradox of Why").

The skeptic would ask, "What is the relative likelihood that the Authority who created earth is lonely and attempting to trick mankind into sharing his burden of eternal life by inviting chosen ones to consent to their inheritance?" In law, consent is naturally a requirement for the transfer of any gift. The English translation of the apocryphal 2 Enoch 24:4–5 reads, "For before all things were visible, I alone used to go about in the invisible things, like the sun from east to west, and from west to east. But even the sun has peace in itself, while I found no peace, because I was creating all things, and I conceived the thought of placing foundations, and of creating visible creation."[13]

The skeptic would continue, "Is it not a common tactic of oppressive regimes to convince their citizens of some greater enemy (i.e., the terrorists), so the nation can justify building-up a 'lesser evil' in the form of a military dictatorship?" Indeed, selling fear appears to be a tactic used by warlike regimes throughout history to build support for their militaries. Did this tactic find its way into the Bible? Is it from a divine source, or from the minds of human warrior societies?

Skeptical questions test our faith, to strengthen us as we craft thoughtful answers. I think we are being tested here on earth, and the most satisfactory answer is found on the narrow path described by the biblical Messiah. I think the previous patriarchs/sovereigns of the Bible (e.g., David, Moses, Abraham) each experienced their own unique, incomplete, and unsatisfactory answers to life's mysteries. In other words, the Bible is incomplete without the Messiah.

As a sinner, why would the creator of earth want to forgive my sins? Is it because I think the creator is powerful and interesting so I pander to him? No. We can start this answer with Matthew 6:14–15, "For if you forgive to sons of men their transgressions, your Father in heaven will also forgive to you. And if you do not forgive to sons of men, also your Father does not forgive to you your transgressions."

In Aramaic, the word forgiveness is routinely synonymous with "release." Why is forgiveness/release moral? Perhaps the answer can be found by analogy—teachers forgive students; parents forgive children; judges show mercy on defendants. One might say: the greater the forgiveness, the greater the love. Luke 7:47, "Her many sins are forgiven because she has loved much; but to him that a little is forgiven, loves little."

Teachers want to forgive students because failure is part of the learning and growth process. Parents want to forgive their children for the same reason teachers forgive, and because it is healthy for families who live together to avoid excessive grudges, and to enjoy new beginnings. And judges

13. Platt, *Forgotten Books*, 131.

offer forgiveness to defendants as a vote of confidence in their reform. See, Revelation 21:5, "Behold, I work anything new."

The skeptic persists, "If my sins are forgiven, am I purchased like a slave? If so, how am I supposed to love my master? Should we both fear and love God as a loving parent (so far so good) who threatens to kills us if we fail to meet the standard of conduct (hmm), regardless of whether that standard appears vague from the human perspective (ouch)." The Christian response here, as expected, is the Messiah's words. In Matthew 7:1–2, 26:26–27 we see that whatever standard you personally use to judge, that is the same standard that will be used to judge you. The golden rule may even be inherently known by all people—do unto others as you would have them do unto you. It is so practical it can be taught to any level of intellect.

Additionally, I find helpful the analogy on earth that superior species (like humans) often live congruously with lower order species (like sheep or strawberries), but will readily extinguish problem members of the lower order species to maintain order. To the extent I think and act spiritually and upright on earth, then I am like a son to my creator, or perhaps a well-behaved student, or maybe even a fig tree. But to the extent I act badly and without reform, then I may be like a pest that risks being destroyed as I draw attention to myself and burden the Creator and his family while they work and play. (For purposes of full disclosure, I must admit that I have killed many aged and young insects that I considered home invaders.)

If a superior being can never prove that he is the omnipotent Authority over everything, then perhaps he cannot fairly require his subjects to completely surrender their choice (free will) to reject one of his orders. Consider the infamous Ezekiel 9:6 command for Israel's army to slay anyone committing abominations in the city, which included women and children.

The words in the Bible are often analogized to food. The Messiah instructed to test fruit to determine if it comes from a good tree. Matthew 7:16–20. That is logical. And he also provided a parable of Satan sowing weeds (obstacles) among wheat (useful), and a rationale for why they should grow together for an age. Matthew 13:24–30.

For the sake of argument, let's assume each one of the marked humans that participated in abominations in Ezekiel 9:4 was like diseased strawberries or perhaps would later go on to monstruous lives. Perhaps a superior being with foreknowledge of this future would be justified in giving the advance kill order. The superior being may even be justified in killing or appropriately punishing any solider who disobeyed the kill order. But, why order soldiers to do the killing in the first place? Is a valuable lesson being learned here about idolatry, or species dominance?

Passages like Ezekiel 9:4–6 are favorites among skeptics to present the possibility that a false choice is being presented: biblically jealous Yahweh or murderous Satan? Do we have all the facts right? Is this some kind of test where we are supposed to select another option?

When should a man surrender his moral compass to the will of a superior being? It seems the values produced would be faith and efficiency. Do these trump the value of choice, even the choice to be wrong?

As things stand, it appears that Yahweh works his own will on planet earth, confusion abounds, and there is no other being that intercedes, because Yahweh complies with his covenants. And ultimately, we are learning a lot through complex sagas. This appears to be a test of morality. Are we simply fans of God because he is powerful, or do we genuinely love his character as the creator of this wild rugged experience? Do we love God's will so much we are willing to become it?

Chapter 4

Justice versus Multiverse

SUMMARY

THIS IS THE MOST difficult chapter for me to write (because I am so passionate about the subject), and the most contemporary because many people are replacing a traditional belief in God with a modern belief in a multiverse. The Bible says God is just, and that anything in God is possible (that is, if God allows it, then it is possible in God). By contrast, people who believe in a multiverse assume that every conceivable possibility exists simultaneously, including infinite gods. One of the immediate costs of their erroneous belief in a multiverse is the abandonment of justice, because if every conceivable possibility exists simultaneously then they believe the multiverse has infinite worlds of infinite injustice.

Relevance for your life . . .	Imagine for your life . . .
You will love God when you understand God is just. You cannot love God if you believe God is unjust.	You love and respect your spouse because you know they are *not* capable of unspeakable depravity. They faithfully and ethically honor your relationship, your family, and the social contract.

Logical Hypothesis

1. If randomness,
2. Then anything,
3. Then probability potential to infinity.

CREATION OF JUSTICE THROUGH FAITH

Faith is a creative force in your head. One possible meaning/translation of the Bible's first word בראשית = "it was in his head".

The Bible says that Yahweh spoke the world into existence. Your brain thinks first, and then you speak, just like the alternate translation of Genesis 1:1, "It was in his head . . ."

God's Oneness creates many paradoxes, and it is essential that God have faith in himself in order to prevent an exponential loophole in the cosmic justice system. You see, when some of us humans observe cause-and-effect laws described in our Bible, we take comfort in our assumption by analogy: that God *also* would not engage in injustice, for consideration/fear that he too would suffer reciprocity for his own injustices at the hands of a higher Power similarly constituted with a universal logic of cause-and-effect retribution. Obviously, we believe there is none higher than the Lord, but it is actually only God's faith in his own Oneness that can close the loophole.

Here is the surprising logic though—the concept of an exponential forgiveness requirement is frightening. Why? Because logically, a law of forgiveness could theoretically be exploited as an exponential loophole in the cosmic justice system to avoid exponential amounts of personal accountability for cause-and-effect. Ask: why would a truly divine being ever need forgiveness; for if they do (or even if they wonder if they do), are they truly divine, and have they not already proven that they are capable of wondering whether they are truly divine? And if they are not truly divine or are wondering so, then perhaps they indeed have the personality in them to engage in torture-testing to explore the Unknown.

So, paradoxically, perhaps the most frightening Authority of all is one who claims to reign alone and creates a universal law of forgiveness. This paradox is resolved by faith that Yahweh is One. Every language (science, math, words) points to faith as the ultimate language (of God). Faith is the only language that is self-fulfilling and therefore necessary. As such, faith is the self-fulfilling description of the first cause, the only thing that cannot be disproven. Faith points to the first cause. What concepts are also self-fulfilling, paradoxical, and require faith?

It is the nature of God that he *is* the thing that he *does*. So, it is inherently controversial to discuss God's power to annihilate because it raises the inevitable point that God is Nothing. This book attempts to raise this point in the most respectful manner toward God, free of all blasphemy to the extent disclaimers can suffice to provide such freedom.

To give eternal life to his son, Yahweh *is* eternal life. So, to give annihilation to his enemies, Yahweh *is* the Nothing.

You (noun) are what you do (verb), so by deduction if there is creation (noun), there is creating (verb).

GOD IS JUST, HIS PUNISHMENTS ARE FINITE AND PROPORTIONAL

In all three of the original Bible languages (Hebrew, Aramaic, and Greek), the literal text reads the same: torment is described exclusively with finite terms.

Consider Psalm 33:5 regarding God, "He loves righteousness and justice." See also Deuteronomy 32:4, Psalm 7:11 ("God is an upright judge"), and Psalm 19:9 ("the judgements of Yahweh are true, altogether righteous"). Because God's way is justice, it makes sense that the ancient literal Bible text says that punishment is real, but *not* everlasting.

Hell is described as *finite* in both space and time. And all punishment is *proportional*, because Yahweh does justice. The text specifies repeatedly that only God and those who live in God (namely, the Messiah) live continually/perpetually. Everlasting hell is simply a false translation, albeit a persistent one.[1]

The Bible may be the best-selling book of all time, but most people spend no time reading the original languages to see what is written there.

The time period of any given punishment is described in terms of an "age" (עלמ in Hebrew and Aramaic, often transliterated *olam*), and "eon" (*aion* in Greek). And in most cases, the punishment is described in even shorter duration, as we see in the many Psalms and Old Testament prophets who spoke of the wicked that are destroyed like chaff (which burns in mere seconds and it is gone).

1. For my deep dive on the ancient Bible text here, see Glaser, "Preacher Man."

If the Bible was going to suggest infinite torment, any of the writers could have done so literally by saying "time that does not end" or "time that cannot end." Or something along those lines. But those words do not appear in the literal Bible anywhere in regards to punishment. Rather, only the Father (and those who live in the Father, namely the Son) are described with special words indicating perpetual life. The fallen do *not* have perpetual life or perpetual torment, but rather sin destroys. Even a permanent death would not translate as everlasting torment.

Accordingly, those who preach infinite punishment or torment do so of their own assumptions and dogma because their assumptions are not supported by the literal Bible. The Messiah invited us to follow his word, not the translator's word. I do not think it is possible to write on a loving heart a desire for infinite torment of creatures.

The Bible instructs that (1) only God lives indefinitely, (2) if you remain in God, you live as he allows you to live, from world/age to world/age, (3) if you depart from God, you die because that which is outside God is eventually destroyed in its season. Humans are not infinite gods themselves; humans are finite in both space and time. The promise of "everlasting life" and "everlasting torment" are both simplistic mistranslations favored by modern preachers. In reality, the literal Hebrew/Aramaic/Greek translation is that your life continues from world to world, age to age, as long as you remain in God; if you ever depart from God's protection you are dead, even if that death occurs seven worlds from this one. You may live from world to world for a while, but there is no explicit literal guarantee in the Bible that you will live for infinite time. See e.g., Daniel 12:2–3, Psalm 45:6.

Indeed, many angels are recorded as starting off in God's protection, but upon transgressing they are captured in chains of darkness (2 Peter 2:4) and scheduled to be utterly destroyed at their judgment day. The text does not claim they are tormented infinitely, but rather they are "destroyed" after their time in prison. In Revelation, Satan too is tormented for a finite time described as "ages" and in a finite space described as "worlds." Ages and worlds are by definition finite in space and time. If God wanted, the literal text of the Bible could have easily said "tormented for time that does not end" or "time that cannot end," but the text does *not* say that. I repeat—the literal text does not say that time continues forever, or is somehow suspended (i.e., outside time). Rather the literal text refers exclusively to finite planetary bodies (worlds) and finite time periods (ages). To claim more than the literal Bible says is to rely on mistranslations. Mistranslations take us away from God.

The English word "hell" is derived from the Hebrew and Aramaic words שאול (*Sheol*: a finite prison or grave) and גהנא (*Gehenna*: a burning trash dump, metaphorically a place of complete destruction by fire), not everlasting existence and torture. For example, regarding Gehenna in Israel, the trash dump fire was eventually quenched so it only burned for a limited time (and limited space), just like Jeremiah declared to the ancient Jews, "And if you will not hear me . . . then will I kindle a fire in its gates, and it shall consume the palaces of Jerusalem, and it shall not be quenched." Jeremiah 17:27. That fire too was eventually quenched because in 2 Chronicles 36:19–21 the literal fulfillment of this prophecy occurred when the Babylonians burned the city. Is that fire still burning in Israel? No. Is the Gehenna trash fire still burning in Israel today? No. Mistranslation enthusiasts would ask, "But if a fire is unquenchable, doesn't it last forever?" No, unquenchable simply means it burns *all of the fuel* given to the fire, such that no one stops the fire before it finishes. So, the Jeremiah "unquenchable fire" was unquenchable because no man was able to extinguish it prematurely. It is that simple to understand once we see the fulfillment of the verse in 2 Chronicles 36:19. So too with the Mark 9:44–48 "unquenchable" fire.

Now granted, it is inherently challenging for us humans to use our simple letters and language to define things we do not yet fully understand, let alone an underworld that we have not seen. But when priests and scholars blatantly mistranslate ancient words to concoct an everlasting torture prison—that is dishonest and evil and unnecessary.

Indeed, many Christians today are surprised to learn it was the Pharisees who preached eternal hell, along with Greek and Egyptian mystics.[2] The original Christians did *not* preach eternal hell (nor did Moses or David or any writer in the Old Testament), but rather the false dogma crept into Christianity especially around the same time that Latin was spread through Catholicism, which was hundreds of years *after* the Messiah was crucified—remember that he warned of wolves in sheep's clothing and false

2. "Nothing is better established in history than that the doctrine of endless punishment, as held by the Christian church in medieval times, was of Egyptian origin and [ ] adopted by the Greeks and Romans." Hanson, *Universalism*, 36–52. I think universalism is incorrect (because the Bible says God saves *any* man, not every man), but the universalist researchers have done an excellent job retracing the ancient origins of hell mistranslation dogmas.

prophets. One of the most pervasive themes in the Bible is the errancy of priests who teach the people false dogma.

Due to obvious later mistranslations in Latin and English of just a handful of words over millennia, most mainstream religious groups have been able to propagate the evil and hateful and false doctrine/prophecy of an everlasting torture prison. Why? Because most Christians are swayed by Bible translations and commentaries above the literal Bible. God probably destined this so we will learn something about ourselves.

Logically, I think God wants Bible readers to be faced with all feasible translations of *olam* and *Gehenna*, etc. for several reasons, such as (1) so that we learn about the nature of fear and love, (2) so that we will take grace seriously, (3) so that we show our true colors, and (4) to provide an extreme illustration of the harm that comes from false doctrines of men.

In my experience, when a person is told that there is an everlasting hell, he or she may exhibit the following negative behaviors: hope that God does not exist, belief that God does not do justice, and preaching the everlasting hell doctrine to others to help share the burden of psychological fear.

This word in Hebrew עלם (*olam*) means "a world beyond, time period, a long time, an age, multiple ages, horizon, out-of-sight." As the Hebrew word requires a context in order to have meaning, the word is protected from priests who try to mistranslate it, especially in a vacuum. Indeed, it is dishonest where modern Bibles claim עלם means "everlasting" or "universe" in only certain preferred contexts to match their theology. Take some comfort in Psalm 12:7 highlighting that ancient scripture is protected. Here with עלם, we see it is protected by context.

Given the seriousness of the matter at issue here, if עלם ever meant "everlasting" (or endless universe for that matter) then there should be *some* statement in scripture defining it clearly as a time period that does not end or that cannot end. I have never found any such statement in the Bible. Moreover, the evidence we have is that biblical persons such as Abraham, Job, Moses, and David did not believe in (or write about) an everlasting torture prison after death. Again, it was certain Egyptian Mystics and Greek Philosophers and Jewish Pharisees who originally preached this doctrine, and it was apparently adopted by the hierarchical Christian church sometime around the second and third centuries.

There are abundant examples showing the term *olam* is limited in space and time. See e.g., Jonah 2:6 ("I went down to the moorings of the mountains. The earth in its bars behind me *olam*; and you raise my life from the pit, Yahweh, my God.") We know from the scriptures that Jonah's experience lasted for precisely three days and three nights (Jonah 1:17; Matthew 12:40), not forever. Jonah's case is important in understanding the true meaning of *olam*. Shut away in complete darkness inside the sea creature, Jonah would have had no means of judging the passing of time. See also Exodus 21:6, Deuteronomy 15:17, and Leviticus 25:46 (if *olam* meant everlasting here, then it would erroneously make one man the everlasting servant of another man). See also, Zephaniah 2:9 and Jeremiah 25:27 showing *olam* as a limited age, because in Jeremiah 49:6 God restores the Ammonites; Psalm 89:37 specifying the moon (a finite astronomical body) endures for *olam*, which is temporary per Revelation 21:1.

There are also some biblical examples where *olam* refers to a past time, and is best translated as "old" or "of old." Genesis 6:4, Deuteronomy 32:7 ("Remember the days of old, discern the years of many generations."), Joshua 24:2, 1 Samuel 27:8, Job 22:15.

When we recognize the finite limitations on a planet's space, we must also recognize the finite limitations on a planet's time. The Book of Revelation and Book of Isaiah confirm the current earth and its heavens will be destroyed in due time—they are both finite in space and finite in time. Revelation 21:1–7; Isaiah 65:17–25. The place of torment, also, is described in the Bible as being a place (a world) that is finite in both space and time; or to say it another way, hell is not everywhere.

In Revelation 21:4, the text refers to the new earth when it says "and death will not be מכיל" in that one particular world/age. Indeed, the key limiting word used in this verse מכיל ("now" or "therefore") is used throughout the gospel in a limiting sense to some measurable amount of time or a measurable conclusion in a current world/age. See e.g., Matthew 26:45 (referring to one night's sleep), Mark 11:14 (referring to the present world in comparison to a world to come), Mark 14:41 (same as Matthew 26:45), John 14:30 (Messiah says he will not speak again at that time rather than ever again), John 17:11 (Messiah says he will not be in the world again at that time rather than ever again). Even the root of the word is מן כול ("from measurement"), or מן ("from") and כל ("any"), so it is naturally dependent on context for meaning (i.e., from who/what/where/when; any of who/what/where/when). Here the context is set in terms of new Jerusalem in the new age/world; so that is the specific where and when that sets the context for how "death will not be now," in new Jerusalem in the new age/world. This verse does not say that the new Jerusalem in the new world/age continues for perpetual time or exists outside of time. To claim perpetual time or eternal time from this verse would be an assumption outside the literal text.

In any case, note that this same statement in Revelation 21:4 is conspicuously absent to describe the dragon's punishment. Rather, the dragon's punishment is explicitly described as "second/dragon death"[3] in the very next passage at Revelation 21:8. After this phrase second/dragon death appears in Revelation 21:8, the dragon is not discussed again in the Book of Revelation, neither in chapter 21 nor in chapter 22. That is because the text just told us that the dragon is destroyed, no longer in existence.

Even if something exists outside the time dimension (eternally), why would we assume it cannot be destroyed from whatever dimensions it happens to rely upon? The Bible says God can utterly destroy soul and body. Matthew 10:28, "And do not fear those who kill the body but the soul they are unable to kill. But fear more he who is able to destroy the soul and the body in Gehenna." See also Matthew 7:13 (destruction) and Matthew 21:41–44 (destroy). And remember Ecclesiastes 12:7, "And the dust returns upon the earth like it was. And the spirit returns to God who gave it." Isaiah 26:14, "The dead do not live. The departed spirits do not rise. Thus, you have punished and destroyed them and perished any memory to them." Isaiah 28:28, "bread is not ground in perpetuity."

It is easy to say "infinite" in Hebrew and other Bible languages. So, the Bible's silence on infinity is very telling about it as a man-made dogma. For example, in Hebrew the phrase for infinite includes "no end" (אינסוף which is the conjugation of אין (no) and סוף (end/bound)). Yet the Bible makes it clear that ages do indeed have an end, as we see for example in Ecclesiastes 3:11, "He made anything beautiful in its time, and indeed gave the age in their hearts, yet so that man might not find what God has made from the beginning even until *the end*."

THE MULTIVERSE PROBLEM: WHERE'S THE JUSTICE?

Hypothetically, if we multiply randomness by infinite possibilities, then we have a multiverse where every possible outcome is manifested somewhere. If there is such a multiverse, then there is no justice because some of those manifested possibilities are unspeakable atrocities. Why bother to be good

3. In Aramaic the words "second" and "dragon" are the same. The wordplay is especially prominent in the Book of Revelation.

in "this universe" if there are infinite other universes where you are currently infinitely raping, infinitely murdering, and infinitely tormenting others with your infinite powers?

The Bible gives the solution to this modern insanity, because the Bible does not say that infinite possibilities exist, but rather possibilities *only* exist in God. So, God is the solution as the great judge, the enforcer of cosmic justice.[4]

And metaphysically, what better way to evidence the limitations on infinity than for God to paradoxically also be, and command, the Nothing?

Is a *hypothetical* injustice any less unjust than a physical one in a universe where all is illusion? What about where everything is faith? It is paradoxical that faith is the most real thing there is, yet even faith may be an illusion at the greatest extent of its intangibility.

Multiverse Science

Consider rocket scientist Wernher von Braun, "I find it as difficult to understand a scientist who does not acknowledge the presence of a superior rationality behind the existence of the universe as it is to comprehend a theologian who would deny the advances of science."[5]

There are some scientists out there (especially cosmologists) who like to say that everything is infinite and the universe is composed of infinite universes (an infinite multiverse). To make this assumption, scientists will leap over regenerative concepts such as tori and fractals, and even ignore fundamental lessons of scaled proportionality and conservation of energy.

Here is an example of one of their old tricks: they split an atom in two, allow the two halves to travel miles apart from one another, and then alter the spin of one of the halves. They correctly observe the change to one-half automatically and instantly changes the spin of the other half, yet some will incorrectly theorize that there must be some "spooky" particle or wave that they cannot observe that travels infinitely fast between them to transfer the information. But there is no observable particle or wave travelling infinitely fast. Scientists have never actually observed anything infinite; they only theorize infinite things.

4. The operative word is "enforce". To effortlessly enforce nothing is to be nothing. The idea that anything requires enforcement to exist is to allow possibility. Ultimately, possibility is required before any force or enforcement (the imbuing of force).

5. Polsley, *Bible and Holographic Universe*, 15.

Consider the concept of proportionality and scales to predict the physical result. It is like when you push one side of a scale down, it instantly lifts the other side of the scale, because they are connected (or entangled) in the same infrastructure. So too with our universe, everything is weighed by the Creator.

Many cosmologists today are unwittingly the new advocates for infinite hell as they theorize a multiverse into existence, where every possible circumstance that can occur does occur. So, for example, a multiverse cosmologist will tell you that there is an alternate universe where everything is the exact same in our universe except that in that alternate universe you injure your eyeball tonight, but in this universe you will not. And there is another alternate universe where instead of injuring your eyeball, you suffer infinite torment (i.e., an artificially intelligent machine from outer space comes to earth and captures you in a biological preservation chamber to keep your brain and body alive for an exponential amount of time while it tortures you). Cosmologists are actually teaching this multiverse dogma at universities today. It utterly rejects the laws of proportionality and justice in the universe. It rejects the fundamental law of conservation of energy. The multiverse dogma is based on infinities that have never actually been measured or observed.

This divorce from rationality should emphasize the importance of being literal and honest with the facts before us, with both the words in our Bible and the physical world we navigate.

Theoretical physicist Paul Steinhardt at Princeton correctly rejected the multiverse idea when he said, "Our universe has a simple, natural structure. The multiverse idea is baroque, unnatural, untestable and, in the end, dangerous to science and society."[6]

See also professor Ivan Karpenko, PhD rejecting the multiverse:

> "It is worth mentioning that the many-worlds interpretation, apart from being hard to prove, comes across another serious difficulty. The very concept of probability loses its sense within such an interpretation. If all possible outcomes are real, why do any of the outcomes become more or less probable? Being statistical by its nature, quantum calculations show that during a repeated experiment, a particle would most likely appear in a certain place, however, the probability of some outcomes may be higher than others. So the particle will not necessarily appear in the most probable place, but it will get there more often. Nevertheless, with

6. Wolchover and Byrne, "Multiverse," para. 11.

> the mandatory execution of all outcomes the meaning of such probability—the foundation of quantum mechanics—vanishes."[7]

My key complaint with any infinite multiverse theory is precisely what mathematician David Hilbert said, "Our principal result is that the infinite is nowhere to be found in reality. It neither exists in nature nor provides a legitimate basis for rational thought."[8]

Indeed, I see nature is inviting us to study perpetual systems in the specific context of sustainability and regeneration, like breathing in and out (or like receiving energy to grow, and transferring energy to seed).

It is hypocritical that multiverse theorists desire exponential bliss yet ignore the exponential torment required of their false theory. The biblical reality is that God is just and his perfect/whole balance is like life, so it consumes itself. Thus, God invites us to be sober, balanced, reasoned, humble, and fair. We are not gluttonous or infinite. We only anger for a short time to balance out wrongs. We are finite beings in finite worlds and ages.

Sense of Humor

Does God have a sense of humor with the multiverse theory? Throughout the history of civilization certain religious zealots have proposed (1) the earth is flat, and (2) there are places "out there" where people are punished infinitely. This was especially prominent thinking during the dark ages of Europe about a thousand years ago. And now today certain cosmologists are telling us something that sounds eerily similar: (1) our universe is mostly flat, and (2) there are places "out there" (parallel universes) where every possibility is expressed (including potentially unspeakable forms of injustice and depravity).

When people sever morality from science, they become more vulnerable to error as living beings—error that nature naturally selects for destruction. I think it is immoral when adults instruct children that they will burn in hell for infinity (i.e., if the child does finite x, or disbelieves y). Today certain multiverse cosmology theorists are attempting with smiles to instruct children that creatures like themselves are *already* suffering infinitely unspeakable torments in other universes because it is the nature of "life."

7. Karpenko, "Concept of the Space Interpretation Problem," 12.

8. Hilbert, "On the Infinite," 201.

Just as certain religious bodies have historically mistranslated and misinterpreted certain words (like Hebrew *olam*, Aramaic *alma*, and Greek *aion*) to an unflinching definition of "everlasting," so too have multiverse proponents potentially misinterpreted "infinity" symbols in their own theorems to mean infinite rather than beyond the currently observable horizon.

In reality, QED is directing us toward concepts like rotational charge, focus, and probability amplitudes. My hope is that this new multiverse-delusion dark ages will be short-lived as nature selects the multiverse for its rightful place next to false translations of everlasting hell. I think accepted scientific methodology will eventually side with those using probability amplitudes to re-categorize infinity symbols as perfectly structured fractals (where charge withstands compression without being destroyed (emerging on the other side of the dimension/other where our tools are not focusing yet), so the wave probability remains intact to retain its natural character).

In that better cosmological framework, an important question becomes whether concepts beyond form *also* require fractalization in order to survive charge compression. This research would gain ground where, for example, phase conjugation explains the charge (energy) that rotates (creating time) and that has inertia filling up space with fields (mass). We should try to calculate the likelihood that a fractal charge results in things we do not consider just in the universe. Indeed, scientists can opine on fairness and justice (beginning with inquiries into conservation of energy, and reciprocal causality).

Newborns preoccupy their minds exclusively with questions of form, movement, and orientation in an instantaneous biofeedback manner. As our conscious reflection evolves as adults, we navigate networks of deeper and calculable principles. We strive to define principles like objective justice and righteousness. And that is a good thing for scientists to do, rather than speculate about infinity.

Consider professor Sorensen explaining how Rene Descartes understood God's relationship with finite beings:

> "If he can prove that God exists, Descartes will prove that we can know the external world. But until God is proved, Descartes can only use the data available to his immediate consciousness. Thus Descartes is confined to a priori proofs of God's existence. Descartes's first proof portrays his idea of God as an infallible trace of God himself. All other sources for the idea are methodically excluded: Descartes could not have obtained that idea from himself because he is finite. Nor can an idea of God result from adding

> two finite ideas (in the way Descartes gets the idea of a unicorn by combining the idea of a horn and the idea of a horse). Adding finite things together only produces finite things. Nor can the idea of God come from subtracting one idea from another. Infinity is not a subtraction from finitude. So the idea of God must have come from God himself."[9]

Multiverse Philosophy

Philosophically, the infinite multiverse theory is also fundamentally dangerous because it strikes these two key words—in God—from this equation: "Anything in God is possible". And the cost of that intellectual proposal (i.e., the infinite multiverse theory without a just God) is the focus of human minds toward believing in infinitely unjust and exponentially torturous universes—where things so unspeakable occur that they would shatter minds into pieces. That is not sustainable for minds.

We must remember, in this universe, the *focus* of the observer is part of the equation. That is a *huge clue* to us as breathing, sentient thinkers who desire to be in harmony with nature and active participants in reality. A scientist who preaches with confidence something exponential that he cannot prove is engaged in something very similar to a priest. As evidenced by man's long history of preaching everlasting hell, there is a danger that real people in this real universe will believe in the multiverse and act upon the belief of moral relativism. It was the belief in everlasting hell that motivated medieval torturers during the dark ages—the torturers thought they were actually helping heretics avoid a worse fate in a *presumed* infinite afterlife. So, history shows a tangible, practical danger from presenting "everlasting torture" logic to sentient beings on earth.

I think the multiverse belongs in science fiction dystopia (notably, Marvel Comics has recently made it popular by romanticizing multiverse dogma). But again, the multiverse does not have a reliable mathematical footing, and it cannot be tested empirically, so it is not appropriate for physicists to present the multiverse as anything other than raw, unadulterated speculation. I would liken it to saying that two parallel lines meet at infinity—that is just raw, un-mathematical speculation. Or it is like saying the value of 0/0 has infinite answers—again, un-mathematical speculation.

9. Sorensen, *Brief History of Paradox*, 218.

Anyone who asserts that absolute truths do not exist is making an absolute statement, which is of course hypocrisy. In a multiverse, absolute truth varies from universe to universe, such that the multiverse claims to be an absolute truth constructed of relative truths.

If we do live inside a holograph, then all the more we are legal fictions like corporations or personalities. We are not supreme gods. We are created beings who owe fidelity to God. Isaiah 44:6, "Thus says Yahweh, king of Israel, and its redeemer, Yahweh of Hosts. I am the first, and I am the last. And apart from me there are no gods."

Living inside an image (and loving the image/world more than the Creator) implicates idolatry.

MULTIVERSE IS A GIGO ILLUSION

How does man claim to understand or see infinity, except by illusion?

Indeed, theoretical physicists manually insert infinities into their equations (simply because calculus requires infinities rather than very large numbers in order to be able to process equations, even with computers), and then voila, they find infinities in their results, and then misleadingly claim that everything is infinite.[10] So the infinities are not based on actual experimental data, but rather upon raw theory and convenience in calculus. That is how multiverse theorists are currently convincing the world to abandon God and morality in favor of a multiverse where every conceivable universe simultaneously exists, including universes with unspeakable depravity. Meanwhile, countless physicists (like Princeton physicist Paul Steinhardt quoted earlier) are admonishing them to stop their infinite multiverse nonsense.

But multiverse physicists/translators preach confidently and do not caveat their work, and the public just eats it up (again, Marvel Comics). So, there is indeed a heavy price to pay when literal words and finite numbers are simply abandoned in man's quest for glory in religion, science, and culture.

Here is another example where an infinity leads to inaccurate results. The difference between "impossible" and "zero probability": "this event

10. Once you assume infinity it is everywhere. It is between 0 and 1 as much as after 1, which is why, "After failing to prove that the set of all real numbers is greater than the set of real numbers between 0 and 1, Cantor surprised himself by demonstrating that all the real numbers could be put into a one-to-one correspondence with the real numbers in the interval between 0 and 1." Sorensen, *Brief History of Paradox*, 323.

indeed has zero probability of happening, but it's still true. This phenomenon is because of the following geometric fact: *it's possible to have a non-empty set with zero 'volume'*.")[11] The mathematicians here (from Duke University) are using the word "true" loosely and inaccurately because of their reliance on infinity. Their example allocates total probability measure one over a finite interval. The interval is a continuum with infinitely many points, each of which is possible, but each of which has probability measure zero because a point has no length. Any sub interval has a positive, non-zero probability, because a sub interval has a nonzero positive valued length.

The example illustrates the idealization of reality by a probability model. This sort of idealization (which invokes a continuum in two or more dimensions) is done often in science, but always the model is only a fictional approximation.

In reality, you cannot physically select a random number in such a way that *all* values within an interval are possible. There are a finite number of particles in the universe, and any number selected by any real-world physical process will be restricted to a finite set of possibilities, each with a nonzero positive probability. For example, you just may not be able to identify let alone choose some number with a gazillion decimal places in a universe with less than a gazillion particles. Such a number has no meaning or accessibility in a practical sense.

So, the notion of a possible event with probability zero is merely the consequence of a mathematical abstraction, a simplification of the real world.

Physicists argue that spacetime is quantized, not continuous, and that all things possible require a ridiculously large but not infinite number. Mathematical models of physics on the other hand employ simplifying equations that involve continuums and necessarily an infinity of possibilities. But the infinity is just a mathematical illusion.

Illusions can be useful, but also dangerous when sold as reality. Irony recognized.

11. Guo et al., "Impossible and Zero Probability," para. 3.

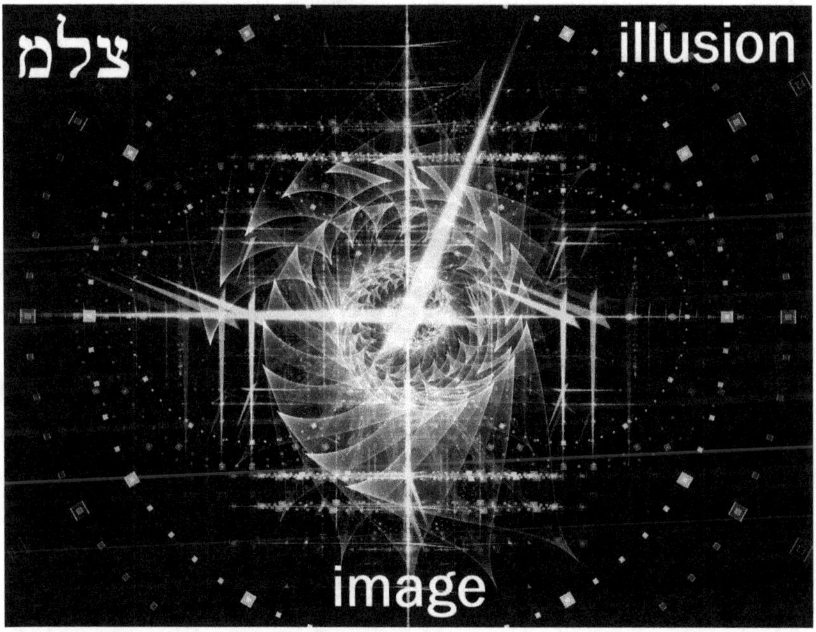

IMPORTANT TIME PROBLEM WITH MY THEORY

In chapter 1, why do I assume time/progression in conceptual Randomness, where shapes *eventually* emerge, rather than assume all possibilities exist simultaneously outside time (i.e., multiverse). Once I assume Randomness, then there is no order and therefore no time. And if there is no time, can there be progression?

This is a problem with my theory in chapter 1 that God emerged *early* from Nothing. How could God know how early or late he emerged?

Was God offended by a hypothetical multiverse-like infinite Randomness from which God emerged, and is this part of the deeper meaning of nakedness? Humans cannot answer. We can only attempt to apply the principle of conservation of energy to the principle of Randomness.

The logical hypothesis for this chapter is, "1. If randomness, 2. Then anything, 3. Then probability potential to infinity." So, if we hypothesize infinite Randomness, why don't infinite gods appear? Fortunately, the answer is obvious: because one infinite God is enough to cover all of infinity.

One could theorize multiple gods emerging from Randomness, but like sperm only one fertilizes the egg and grows fractally into a baby. Like so much of this subject, one can only speculate. We need God to educate us. And God is teaching us now to find him through his son the Messiah, who is good and just.

Chapter 5

Paradoxes of Law and Free Will

SUMMARY

Paradoxically, we feel most free in a society of law and order that restricts us in some measure. This measurement teaches us our character, in the context of our values. Biblical and legal history provide excellent and often surprising examples of the social contracts we form, break, and reform on this rugged planet.

Relevance for your life . . .	Imagine for your life . . .
Law provides order to society and your life. When you respect the law, the law respects you.	When your family follows the rules, everyone feels more free and respected.

Logical Hypothesis

1. If limits exist,
2. Then law exists,
3. Then a lawmaker exists.

LAWS OF ONE AND NONE

The Bible refers repeatedly to the words for *all* and *nothing*, and there are many different word choices to convey these concepts. The same is true in English as we have multiple words for the concept *all* (every, each, total, any, complete, etc.) and the concept *nothing* (none, no, zero, void, absent, etc.).

Recall John 1:3, "Everything in his hand existed and without him not even one existed whatever that was."

The most famous "law of one" is the Greatest Commandment, "Hear, Israel: יהוה our God, יהוה is one. And you shall love יהוה your God with any/all of your heart, with any/all of your soul, and with any/all of your strength." Deuteronomy 6:4–5. This is followed by the second commandment, which people recognize as a continuation of the "law of one" and indeed it reminds of quantum entanglement, "And the second [commandment] is like it, that you should love your neighbor as yourself." Mark 12:31. See also John 13:34 "A new commandment I give to you, that you be loving to one another. Just as I have loved you, also you should love one another. In this, any man will know that you are my disciples, if there will be love in you."

According to the Jewish Talmud (Shabbat 31a), an unbeliever once (mockingly) asked the famous first century rabbi Hillel to teach him the entire Torah while the unbeliever stood upon one foot. Hillel replied, "What is hateful to you, do not do to your neighbor. That is the whole Torah; the rest is the explanation of this—go and study." The rabbi's response was clever and it illustrates the importance of cleverness to the Jewish people when the Messiah taught among them at the same time as Hillel. And indeed, the Messiah explained the Torah better to those *on two feet*, in Matthew 7:12, "Whatever you want men to do to you, do also to them, for this is *the law* and *the prophets*." In the analogy, the law is the first foot, and the prophets the second.

A prophecy has many ways to come true. See e.g., Luke 19:40. This helps explain the difference between free will and fate. Humans appear to have the capacity to exercise free will within a limited spectrum of possibilities, but ultimate fates are determined by God who commands possibilities. Our limited spectrum is still quite vast though. See e.g., John 10:34 (where the Messiah refers to men as godlike, "Is it not so written in your law: 'I have said, you are of God?'"); Genesis 1:26 ("And God said, 'Let us make man in our image.'"); Matthew 17:20 ("If you had faith in you like a mustard seed, you could say to this mountain, 'depart from here' and it would depart, and nothing would be able to prevail over you.")

God's immense and unrivaled power is a consistent theme in the Bible. See e.g., Exodus 20:3 ("You shall have no other gods before me."); Luke 1:37 ("Nothing is difficult to God."); Jeremiah 32:17 ("Ah, Lord יהוה. Behold, you have made the heavens and the earth by your great power and outstretched arm. There is nothing too hard for you."); Isaiah 55:8 ("For my thoughts are not your thoughts. Nor are my ways your ways, says יהוה.")

We learn that God's law is truth and righteousness. See e.g., Psalm 19:7–11, 119:142–51. And indeed, there is some biblical law that continues today, according to Matthew 5:18.

ALL OR ANY

It bears repeating here the translation "any" or "all" in biblical Hebrew and Aramaic. When a humble Israelite says כל he is literally saying "any" in the sense that "any" could mean "all" but it first means "any." This is very important because it is the difference between "Christ will save *any* man" compared to "Christ will save *every* man." The actual literal translation in the ancient text is "any." If God ultimately chooses to save every man, technically that would still be "any" man (thus the literal translation remains true), but it will ultimately be God's choice, as the humble Israelite knows. The word כל does not categorically mean "all," so there is an essential humility and honesty in the literal reading "any." Indeed, in the first century, if an Israelite wanted to clearly say "all," he would say both the affirmative and the negative together (i.e., "please pick up כל ("any") apples and do not leave even one on the ground"). See e.g., Psalm 34:20 ("And he will keep any of his bones. Not even one of them will be broken."); Psalm 14:3; Joshua 23:14; 2 Samuel 13:30; John 1:3.

DEFINING MURDER

In the gospel of John, we learn that Satan (the blamer) is a liar and murderer. In chapter 7 we will discuss the equivalence of lying and blame (where blame is the absence of self-honesty). But first we will discuss murder, to prove how it relates to blame.

One of the ten commandments is "You shall not murder." It is paradoxical of course that the murder of the Messiah culminated in the great sacrifice restoring life. Why is God using paradoxes to teach us about himself?

What is murder and who is capable of this powerful force? For example, when a soldier kills another solider in battle, we typically do not call it murder. We do not blame the soldier because it is called a justified killing. And its roots are biblical.

In the gospel, the word for "kill" קטל is slightly different than "murder" קטול.

Dictionaries like Jastrow and Payne Smith note this, at least in part.[1]

But the vav-placement is not a hard rule for definition as "murder" because conjugation alone cannot bring about the legal conclusion "murder." In the gospel, you need a judge or accusation to do that, which is why we see the conjugation קטולא ("murderer(s)") only when the judge appears in Matthew 22:7, but not earlier in Matthew 22:4–6 (where we see קטל ("kill")). See also John 16:2 (showing temporal retroactivity); Revelation 22:15; John 10:10 (already judged a "thief"). This is a subtle point in the gospel, but an important point that can be easily overlooked if we judge killers prematurely. Matthew 5:21, "You have heard that it was said previously: לא תקטול ("you should not murder/kill"). See also Mark 10:19 קטול ("murder"). After this, the Messiah describes that murder is really a matter of the heart. And see for example, who will judge the heart, per 1 Samuel 16:7 and Matthew 19:18 תקטול ("you murder/kill").

So, "murder" is a legal conclusion, not a translation of a single Hebrew or Aramaic word into English. It is also a question of righteousness and a matter of the heart. If the killing is done by God, even the strongest expression of קטל ("kill") in Aramaic קטול ("kill") would not translate murder; it just means premeditated killing. If a killing is righteous or forgivable, it is not murder, it is just a righteous or forgivable intentional killing.

Second, context is an important clue in Aramaic to see the distinction between "kill" and "murder," but sometimes our assumptions about context mislead us because we are imperfect and premature judges. The conjugation can be quite surprising in Aramaic. See e.g., Matthew 2:16 where Herod קטל ("kills") infants rather than קטול ("murders"), perhaps because of the legal doctrine of sovereign immunity;[2] Matthew 14:5 (the crowd thinks they are righteous to desire to קטל "kill" Yahshua because they deny he is the Messiah); John 8:11 ("suicide" not pre-judged as murder); John 8:33 (Satan קטל "killed" men in the beginning).

Also, in Aramaic when a person is described as *being* "killed" the expression is קטלת ("killed") or similar, so you cannot tell by conjugation

1. Smith, *Syriac Dictionary*, 500; Jastrow, *Dictionary of the Targumim*, 1346.

2. The doctrine of sovereign immunity specifies that the king can do no wrong. How is that for righteousness?! Sovereign immunity is consistently recognized as current law in the world today (which is why in America, for example, we have the Tort Claims Act where the government *consents* to allow lawsuits against it). See also, Flatto, "The King and I," 98.

alone if the suggestion is "murder". Rather, you need to examine the killer and context to learn if it was murder. Whenever the Messiah described how he was going to be killed, he never prejudged it murder or anyone a murderer, but rather the passive conjugations are used consistently in each gospel. This emphasizes an important point about the *timing* of judgment.

Third, a key point of clarification comes from considering this verb קטול ("murder," "kill") as a type of imperative form of קטל. Usually, the verb imperative is a command (i.e., "go there," "do this"), but in the bigger picture it is about *expressed intention*; thus, the heart.

Unlike Aramaic, Hebrew does not form prohibitions with the imperative, so for scholars historically these conjugations added to the ambiguity (i.e., the voluminous discussion about the Hebrew imperfect tense and vowel points in the Old Testament regarding this very question of murder versus killing).

Here is the subtlety: if a man takes it upon his own mind to kill, that is like an imperative קטול (i.e., he is ordering the killing). By contrast, if you kill in self-defense (i.e., you kill an attacker to survive his attack) you would not use the imperative-style conjugation about yourself (i.e., your heart is not ordering the killing, but rather circumstances dictate the killing; or in other words, you were forced to do it). The same rationale applies loosely to non-human killings, such as the act of killing plants and animals to eat. These are all interesting and often rhetorical questions. For example, what about sacrificial killings (to "defend" your soul, to "restore righteousness")?

Caesar's chief title in Latin was *Imperator* ("Emperor"), and it is from the same root word as "imperative" referring to the language tense. This classic example informs our understanding about the greater meaning of the imperative tense in our world built upon words—orders come from the *Imperator* in the imperative. Who is your *Imperator*?

The imperative tense is about self-determinism. God is the Supreme Commander of the imperative tense.

If someone (like Caesar) orders you to kill, then you are not acting in the imperative yourself (you are a soldier, not a murderer). Of course, things get messy with killings, and answers are not always clear. In the modern era we have the Uniform Code of Military Justice 809 Art. 90, specifying that military personnel are only asked to obey a "lawful" command or order of a superior officer. And then there is the famous Latin expression *Lex Injusta No Est Lex* ("an unjust law is not a law").

In the meantime, who shall judge the killings? See Revelation 6:9–11,

> "And when he opens the fifth seal, I saw toward underneath the sacrifice altar, to the souls דאתקטלי ("who were killed") because of the word of God and because of the testimony of Yahshua that they had to them. And they cried out in a great voice and said, 'Until when Lord, holy and true? You do not judge and you do not avenge our blood from the inhabitants of the earth.' And white garments were given to anyone from them. And it was said that they will be resting until a little time, until their fellow brothers have also been delivered, those who are preparing to be killed like they also were."

In summary, the lesson is to not say murder unless you intend to judge the killing.

THE LAW OF WORDS

Here is an Aramaic literal translation of John 1:1–2, "In the beginning was his existence, the Word. And he the Word had his existence toward God. And God was his existence, he the Word. This was his existence in the beginning toward God."

There are five occurrences of איתוהי הוא by the time we get to John 1:9 (and one plural reference of same in John 1:4), and the concepts that this phrase connects are very mighty and spiritual: God, light, darkness, Word, life. And this emphasizes a key point: the entire gospel provides the saga of real people in real life situations to help us understand the relative meaning of the Word to our perspectives. This reminds me of relativity as a governing principle of physics.

The Word is very complex in the sense of the fundamental descriptions in John 1:1–2, such as God "has his being toward" the Word. In an energetic sense, possession involves coupling (i.e., electromagnetically, gravitationally). In a logical human sense, אית can refer to a recognized ownership or personal bond between two objects or people (i.e., father and son, or water and cup). And in a directional sense, לות means that a person/object is pointing toward another person/object, and its use in the gospel can imply an energetic connection between the people/objects, or the purpose of the pointing person/object.

I reason that Yahweh is the Word (John 12:48–50) that spoke light into existence per Genesis 1:3. The Messiah is the light created by the Word and living in the Word—that is a key message of John 1:1–9, and passages like John 16:15. We are invited to be sons of Yahweh by following the light toward the Word. In Aramaic, מלתא ("the Word") is the beginning and the end of our entire existence here: מלתא is the full (מל) sign (תא) of our existence. We are therefore characters in a living story, and we are completely and wholly dependent on Yahweh for this ongoing gift of life/saga that is destined for fulfillment. Like sound (i.e., words), light works on wave principles. Light and letters are similar in the ability to divide and create and describe reality.

In Aramaic the words for "revelation" and "wave" are the same: גלא. If you are in tune with Yahshua the light, then his story resonates in your heart and mind, like awesome song waves you love to hear, like beautiful color waves you love to see, like words you love to follow because you love light and not darkness. When we do not follow Yahshua, then we do not really love him fully by keeping the Father's word. John 14:24. But if you love Yahshua, he will lead you to Yahweh because Yahweh loves Yahshua, and as Yahshua's follower you live in Yahshua the light, so Yahweh loves you too. John 14:21. Yahweh desires for his light and Word to dwell inside the ones he loves. John 14:20–23. But before we get to that point of happy new beginnings in a new age/world, our story needs to involve judgment of this age/world by the Word, which is Yahweh. John 12:48–50.

These are fundamental principles in Christianity that define and provide context for law and judgement. Law is necessary because of conflict and division, which are concepts emphasized repeatedly in the book of Genesis, setting the tone for the whole Bible.

WORDPLAYS

In biblical Aramaic, there are abundant wordplays throughout the gospel (multiple wordplays on every page), which convey deeper meaning to verses.[3] They also provide a certain measure of strong evidence supporting Aramaic primacy (the theory that the gospel was originally written contemporaneously in this native language of Israel spoken by Yahshua and the apostles, rather than decades later in the foreign language of Greek where Bible translations were in populist demand). I often wish I had the opportunity to sit down with Greek enthusiasts and read the Aramaic gospel, so I could share exciting wordplays on every page of the ancient codex in this language of first century Israel.

3. Glaser, "Wordplays in John"; "Wordplays and Segways"; "Abundant Wordplays."

One of the first wordplays comes from מנהר in John 1:5, "And that light in the darkness shines and the darkness not did overtake it." This Aramaic word מנהר ("shines") has an alternate literal translation meaning "from fight" (מן = from; הר = fight), as in the very battle of light and dark energies being described in John 1:5. So through *division* of this one word, we see a complimentary meaning in John 1:4–5, because the "light of men" in John 1:4 was divided/separated just like light and dark were separated in the beginning. Is this one of the deeper reasons that certain angels (beings of light) were separated from heaven? Recall that Lucifer fell from grace and eventually receives "chains of darkness." 2 Peter 2:4.

How was the light of men separated? Adam was one—he was man, not a son of man like all men after him. After his second and third separations, Cain and Abel, his separations competed with one another. At first it appeared that darkness (the stumbling block, the shadow) triumphed when Cain murdered Abel, but then Seth was born, and eventually the Messiah was born from Seth's line, and so "the light of men" triumphed; John 1:5 is a true statement, "And that light in the darkness shines, and the darkness did not overtake it." See also John 14:30, "After this I will not speak much with you, for the ruler of this world comes *and in me he does not have to him anything.*"

The second wordplay in John 1:5 comes from the word אדרכה ("did overtake it"). This plays on two words: דרכתא ("threshing") and אדרא ("threshing floor"). The metaphor is light separated from dark, like wheat separated from chaff (by beating it with wood (remember the cross) on the threshing floor). Between light and dark, only one has the seed. Light has the seed; wheat has the seed. The darkness, the chaff, is just the cover. The light/seed must break through the darkness/chaff. So, our earth is like the threshing floor for saga, which is a metaphor explaining why our existence here must be filled with such tribulation. Apparently, we profit from getting the seed beaten out of us. Not all seed must be beaten to extraction, but some seed is.

And indeed, nature is full of these examples (even the classic—chicken and egg "paradox") where something living must break a shell to grow, and if it does not first break its shell, then (1) it will not have been protected during its infancy, and (2) it will not be strong enough to face reality outside the shell. Consider the extraordinary power of high frequency light—for example, if we even gaze at the sun for one hundred seconds straight, we can go blind. How much more if we look upon God's light before we have broken our shell and left this existence (which has been designed to make something within us stronger).

The physics of dark energy is particularly mysterious. Light is the electromagnetic radiation of *any* wavelength, even though we laymen tend to define the word in terms of the light we see with our eyes (the visible spectrum). But in the gospel Yahshua is the physicist, constantly emphasizing the importance of the powers that man cannot see. See e.g., John 1:18, John 3:8, John 18:36, and especially John 16:12 ("Again, I have much to say to you, but you are not able to hold now.") In that last verse the word is אחד, meaning "hold" or "seize," which I think suggests that in these human vessels we cannot contain or control other waves and dimensions, to be revealed later.

See also the interesting reference in John 16:25 to עין בגלא ("eyes in waves"), as Yahshua describes how he will speak of the Father in the future, rather than in comparisons/parables. Here is a standard translation though of John 16:25, "These I have spoken with you in parables. But the hour comes when I will not speak to you in parables, but I will *openly* (עין בגלא) make known to you about the Father."

To the extent the Father and Yahshua can be understood in some way by studying the physics of light, examples are fun to ponder. Consider how white light stimulates all *three* cone cell types in the human eye, which is why our eyes see the color white as *all* the wavelengths of the visible spectrum. In the Bible, it appears that one of the primary meanings of wearing white clothes is acceptance into the kingdom of heaven. See e.g., Revelation 7:14, Matthew 17:2.

With that in mind, remember Matthew 19:14—after children come to Yahshua and he approves in reference to the kingdom of heaven, in the next passage (at Matthew 19:16) a man comes to Yahshua and asks what he must do to enter the kingdom of heaven, and Yahshua advises him to keep the commandments, and further that if the man wants to be mature and to follow Yahshua and to have treasure in heaven then he should sell his possessions and give them to the poor. While the disciples are worrying about the seeming burden of these words, Yahshua adds to them in Matthew 19:24, "And again I say to you, it is easier for a rope to enter in the eye (חרורא) of a needle than a rich man to enter into the kingdom of God."

Regarding white (חור) light, note the wordplay here in Matthew 19:24 "the eye" (חרורא) followed by Matthew 19:26 "looked" (חר). I think the meaning here is that if disciples seek freedom, they are going to need some white clothes. Mark 9:3. They must whiten themselves to enter the kingdom of heaven. Scientifically "white" is a light wave perceived by optical nerves (or measurable in a laboratory as the full visible spectrum within the electromagnetic spectrum), but just imagine how rich חור the experience if the eye saw spiritually, beyond human optical nerves. Perhaps that is part of the larger lesson here—there is more to life than the visible spectrum. There is more to being born from the beginning than just another earthly experience.

In John 8:20–21 (truth will set you free), the phrase "will set you free" (נחררכון) comes from this root word meaning "freedom" or "the eye" (חרורא). And the rope (גמלא) from Matthew 19:24 is a collection of twisted fibers (גלא), an Aramaic word that is also translated as "wave," "straw," or "reveal" (see e.g., Matthew 7:4).

I reason that some darkness does not comprehend light because it is a stumbling block, a cluster of obstacles that block the transmission of light. So, the wordplay of John 1:5 fits well—conflict/fighting tends to disrupt/block harmonic wave signals.

THREE TESTS

In John 16:8–11, Yahshua describes the three tests of the Holy Spirit. Each one requires a successively more challenging logical deduction, and the second and third are so challenging they come with their own Aramaic wordplays to suggest the answer to the listener.

First, the Holy Spirit reproves of sin because of a person's unbelief in Yahshua. The deduction is natural—Yahshua is the sacrificial lamb who takes away the sins of those who believe in him.

The second wordplay requires a little Hebrew knowledge and the book of Acts. The Holy Spirit reproves of righteousness because Yahshua goes to the Father and these disciples will not see him again (they will not see him after Pentecost that is, because that is when the Paraclete will be given). The deduction here is regarding the word זדיקותא ("righteousness") because the Hebrew root word צדק means "upright" or "straight." Accordingly, through wordplay Yahshua says he is literally going straight up (upright) directly to heaven to the Father, and the disciples will look on earth (for righteousness, for Yahshua) but not find him on earth. *Note*: this point about Pentecost is further clarified in John 16:16, where Yahshua specifies that they will only see him again for a little time after the crucifixion, but again, these disciples will not see him again after Pentecost. That is a fulfilled prophecy.

The third wordplay may change the way some readers see the world. On the surface, the text appears to say the Holy Spirit reproves of judgment because the ruler of this world is judged. But first notice the triple wordplay here with דינא ("judgment"), דין ("and"), and דין ("judge/law"). Next, notice the curious grammar that concludes the sentence דין הו (alternate translation: "a judge/law he is"). The wordplay is that the ruler of this world is a judge who will be judged. The ruler of this world *is the law*! Kings and kingdoms have come and gone. Men and women have come and gone. The law has governed their very rise and fall and it continues to do so in this moment; it both empowers and destroys. פרקלטא ("Paraclete" referring to Holy Spirit) literally means "break the curse". And just what is that curse that causes death? The law.

I think this answer has been right in front our eyes but we do not see clearly. Indeed, recall the second test, where the Messiah says the Holy Spirit reproves of righteousness because Yahshua goes to the Father and the disciples will not see him again (they will not see him after Pentecost that is, because that is when the Holy Spirit will be given). We need Spirit to see clearly.

This point about Pentecost is further clarified in John 16:16, where Yahshua specifies that the disciples will only see him again for a little time after the crucifixion. Because that is a fulfilled prophecy, beware anyone in history who claimed to have seen Yahshua after Pentecost. Indeed, this is why the author of the Book of Revelation never actually saw the Messiah directly on earth, but rather only saw *an angel* sent by the Messiah. Revelation 22:8.

In John 20:22–24 (when certain apostles receive Holy Spirit), the apostle Thomas was *not* present to receive the Holy Spirit, which explains why the John 14:15–18 and John 15:26 prophecies (where the Messiah says to *all* apostles except Judas Iscariot that they would eventually receive Spirit) was not fulfilled until Pentecost (when Thomas would have finally received the Holy Spirit).

This presents a huge problem for modern Christianity because in the book of Acts, Paul says he saw the Messiah *after* Pentecost, and he says the experience happened on the road to Damascus (which is probably in the wilderness). So, Paul's story presents a potential conflict with the gospel of John (apostles will not see Yahshua after Pentecost) and Matthew 24:26 (if someone says they see the Messiah in the wilderness do not believe it). Such a conflict is not a small matter, especially because the Messiah's chosen apostles (from his chosen twelve) had their own issues/disputes with Paul, in part because Paul never walked with Yahshua yet claimed to be an apostle by virtue of a personal revelation. For starters, I think all Christians can agree that Paul is not necessary to anyone's salvation. Only Christ is necessary to salvation. Many Christians today have raised serious questions about the weight of Paul's words; for example:

1. Paul came late on the scene and never walked with Yahshua. Paul was embroiled in some ongoing conflict among the early Christians. See e.g., Acts 15:2, Galatians 2:11–14. Was Paul rejected by the assembly at Ephesus? Acts 19, 2 Timothy 1:15. If so, that is concerning because in Revelation 2:2, the Church of Ephesus is praised by the Messiah for rejecting "those who claim to be apostles, and they are not". Paul is the only person on record self-claiming to be an apostle and self-claiming to be rejected by Ephesians. So, if the Messiah was not talking about Paul, then who was he talking about? It is a fair question.
2. Why did Paul appear to change and contradict his own conversion story, expanding on it in more detail once he started growing in popularity/influence? Compare Acts 9:3–6 (short version early in time) to Acts 26:12–18 (expanded version later in time). Likewise, Paul claims he saw Jesus of Nazareth on the road to Damascus, which was likely *in the wilderness*. Even if the road was not in the wilderness, it was still not a vision "like lightning visible from east to west" as the Bible would appear to require for a second coming after Pentecost. Thus, Paul's story appears to conflict with the Messiah's warning in Matthew 24:26–27 regarding messianic imposters, "If therefore they will say to you, 'Behold, he is in the wilderness', you should not go out, or 'Behold, he is in an inner room', do not believe it. For just as lightning goes out from the east and appears unto the west, so will the coming of The Son of Man be."
3. Why did Paul use spiritual power to blind a man in Acts 13:11? The Messiah and his chosen apostles never used spiritual power to blind anyone. Every Bible book (gospels and letters) about the apostles who walked with Yahshua shows these apostles were harmless as lambs, excepting (1) the incident where Peter cuts off the ear of the high priest's servant, and the Messiah immediately admonishes Peter and heals the servant, and (2) Acts 5:1–11 believed to be written by an affiliate of Paul. What spirit empowered Paul to blind a man rather

than heal him? It's a fair question. The miracles attributed to Paul read suspiciously, given that they are not mentioned in Paul's letters sent in real time to people who could verify Paul's statements in real time, but rather the attributed miracles are stated in the Book of Acts that was written by Paul's affiliate potentially decades after the events in question, thereby removing the likelihood of actual witnesses verifying or disputing the events. And indeed, there is a complete absence of corroborating percipient witnesses (i.e., other texts contemporary to the events). Some of the attributed miracles, if true, would indeed be powerful, such as the exorcism in Acts 16:18. But others are less impressive. For example, in Acts 20:9–10, a man falls from a balcony and his state of consciousness is unclear when Paul revives him. Is that a miracle, or just a natural course of events in time? It can be medically normal for a person to be unconscious only for seconds or minutes and then to naturally revive. In Acts 28:3–6, when Paul was bitten by a snake believed by locals to be venomous, the text says he suffered no disfigurement—again, the same thing happens to people often with snake bites. It can be medically normal, either because the snake did not release poison or because the body does not disfigure from the quantity of venom released. Paul apparently suffered personal ailments in his life that he could not heal and from which he did not recover. Galatians 4:13–14.

4. Why do churches today call Paul's writing "scripture," when the literal text of the Bible plainly calls them just "writing." For example, how is this statement in 2 Timothy 4:13 inspired, "When you come, bring the cloak that I left with Carpus at Troas, and my scrolls, especially the parchments"? I do not think it would be appropriate to imply "So sayeth the Lord" to Paul's words. The word for true scripture is "gospel" or "law." Peter never referred to Paul's writing as gospel or law.

Personally, I think that Paul said many beautiful and important things, but I would take the same position as the apostle Peter that Paul's writings are just writings, and not gospel. My intention here is not to cut off any ears, but to respect them.

MAXIMS OF JURISPRUDENCE

Maxims of jurisprudence are philosophical statements on the law books but are not binding law. They are statements in the civil code that assist litigants and courts in administering justice. A maxim does not technically change other written laws, but rather a maxim serves as a general principle to aid the just interpretation of laws.

Here are some examples from the California Civil Code:

> 3510. When the reason of a rule ceases, so should the rule itself.
>
> 3520. No one should suffer by the act of another.
>
> 3521. He who takes the benefit must bear the burden.
>
> 3523. For every wrong there is a remedy.
>
> 3530. That which does not appear to exist is to be regarded as if it did not exist.
>
> 3531. The law never requires impossibilities.
>
> 3538. That is certain which can be made certain.
>
> 3542. Interpretation must be reasonable.

Each of these maxims presents a general truism and has a paradoxical quality. For example, consider the maxim at section 3510, "When the reason of a rule ceases, so should the rule itself." It implies the reason and the rule are actually *the same thing*.

This logic also applies to the maxim at section 3542, "Interpretation must be reasonable." It implies the interpretation *is* reason itself. And again for section 3521, "He who takes the benefit must bear the burden." The implication is the benefit and burden are the same thing. Yet again this logic applies to section 3523, "For every wrong there is a remedy." Can you easily imagine a wrong and a remedy that are the same thing, such as when something "wrong" happens to you but it ultimately helped you—what if these are not isolated incidents but rather are the norm once we have sufficient perspective to see God's plan?

This thinking challenges our view of morality, as does the maxim at section 3520, "No one should suffer by the act of another." How could this maxim ever be satisfied? In chapter 7 we explore the premise that God's universe is perfectly designed and is 100 percent fair (the Bible says repeatedly God is perfect and just). What if *all* suffering is self-inflicted when understood in a quantum context that our self-honesty is unstoppable?[4] In such a paradigm, not only would the so-called victim and aggressor be the same person, but there would really be no such thing as ultimate victimhood in this world where everything is opportunity, where everything is quantum connected, where God always delivers justice ultimately.

The maxim at section 3530 works quite well with paradoxes about Nothing, "That which does not appear to exist is to be regarded as if it did not exist." It is also about what we know versus what we do not know. We know inherently we are supposed to do unto others as we would have them do unto us, and we are to do good. We know we are supposed to place responsibility at the source of harm with rational causal justification for judgment. We know we are supposed to create value, and not falsity or error.

Given that God is inherently unknowable, we naturally fear or revere God and afford others the right to do so as well. Theoretical honesty is when you admit your unknowingness, and you allow explanations of life to be addressed through probability. God created probability to help us know truth or the means to it.

This is the religious context of the social contract we have with each other, where we theoretically agree on truth by embracing causality and promoting the good of the golden rule. Man has this social contract with the populace and his environment.

OATHING YES OR NO

Here is a riddle: where are you if "yes" means "no," and "no" means "if," but "yes" does not mean "if" unless you speak the common language of the people and acknowledge that you cannot control any of these answers?

If you speak biblical Hebrew and Aramaic, you can find the answer in Matthew 5:37, and here are some helpful definitions. In biblical Hebrew אין means "no." But in biblical Aramaic (spoken in the same time and place) this same word אין means "yes." And in biblical Aramaic לא is "no." But in biblical Hebrew לא can mean "if." Exact pronunciation and context were important. And then there is the Aramaic ולא with two quite conflicting definitions: "and not" or "it is proper."

4. Recall our observation earlier in the book that the unstoppable force *is* the immovable object. This solves the paradox. In other words, imagine your self-honesty is quantum entangled with the entire universe. You cannot escape self-honesty any more than you can escape the universe.

This situation in first century Israel created many confusing circumstances for those present. It helps put into perspective the Messiah's words in Matthew 5:37, "But your word should be yes, yes (אין אין) and no, no (ולא לא). Anything that adds to these is from evil."

The context of Matthew 5:37 is the Messiah advising his disciples to avoid oaths and swearing by things they cannot control (see Matthew 5:33–36). In ancient Israel there was an eroding tradition for performing certain vows and oaths only in the Hebrew language. See e.g., "There was a rule to the effect that oaths must always be taken in Hebrew (Yad, Shevu'ot 11:8), but it was later mitigated so as to allow the oath to be taken in the language best understood by the deponent (ibid., 11:14; Sh. Ar., HM 87:20)."[5]

Are we still living in Babylon today? אין

So of course, the answer to the riddle is Israel in the first century AD. Consider also the next verse Matthew 5:38 ("You have heard that it was said an eye in exchange for an eye and a tooth in exchange for a tooth.") Notice the wordplay of אין (yes) and אינא (eye). Also, שן (tooth) can be translated as "go away," which has a logical interplay with לא (no), the opposite of אין (yes). Note also the Matthew 5:38 wordplay of tooth (שנא) with hate (סנא), which helps provide a transition to Matthew 5:39, commonly translated "do not oppose an evil one," but the literal is "do not rise to oppose in evil." The literal meaning comports with the gospel message to not exchange evil for evil, but rather good for evil. Yahshua provides literally the best answer.

Also consider this greater point of Matthew 5:21—the example is about burning passion toward death, rather than correcting assumptions about what you have and what you do not have or lose. Thus, continuing the original point, we find the word "cheek" (פכך) in Matthew 5:39 ("But he who strikes you upon your right cheek, turn to him also the other"), which is a play on the word "dull/fool" (פכה). But first, a slap on the cheek is just a facial injury. In Matthew 5:39 the word "also" (אפ) is a play on the word for "face" (אפ) because the injury is just a surface injury. And the whole point of these examples in Matthew is seeing beyond the surface/face.

Back to the word "no" (לא), consider the Matthew 5:31 wordplay of divorce (דוללא) with Matthew 5:22 fool (ללא). In this context, the Aramaic word "fool" (ללא) and the Hebrew word "if" (לא) are a play on one another, and one of the greater contexts to explore here is separation. Remembering

5. Bard, "Oath," 5. This source also points out the close link between the Hebrew words for curse and oath, where these words are sometimes used interchangeably in the Bible. See also, Cartledge, *Vows*, 26–35.

back to Genesis 1:4, how light and dark were separated—the physics of light and dark is largely explained by probabilities which often seem counterintuitive[6] even in the most sophisticated laboratories—another reason to be careful not to call any brother a fool (ללא). Matthew 5:22–24.

The dichotomy of words in Semitic language reveals part of the *character* of Yahweh. For example, with a single action Yahweh can *destroy* one person and *liberate* another—a single word with dual meaning ("destroy/ liberate") helps convey that reality.

In Hebrew, one of the most fascinating dualities is present in the word אל. As stated in chapter 1, it means both "no" and "God". It also works as a participle meaning "in" and "at." The duality of אל ("God / no") represents the reality that God *gives* and God *takes away*. See Psalm 27:9. Anything in God is possible, including negation. As the Lawgiver, God is the one who says "no." As the Creator, God is the one who says "yes."

Yahshua helps reveal this quality of God in his name ישוע. Without the י, his name שוע can be translated "false," but with the י (yod), his name is ישוע ("salvation"). Who is our salvation? Answer: יהוה. What are we without Yahweh? Answer: false.

In Hebrew, a routinely used name for God is אלהא pronounced where אל = God; לא = no; הא = behold.

THE LAW OF PERSPECTIVE

Yahweh sent his unique son, the Messiah, to this world to fulfill the law. Matthew 5:17. Was Yahshua referring to the law of God, or the law imposed upon the people by their rulers? Both? Whose law required death?

Yahweh gave the builders/rulers the power to set law, and yet he also gave them his law. Was conflict inevitable? I think so. Therefore, we ultimately needed justice and forgiveness to have a reconciliation. But in order to receive the virtues of justice and forgiveness, we humans first like to feel *trust* in a relationship with those judging us and forgiving us; it is part of our character to generate trust. In this subsection on perspective, I will especially share my personal perspective. When Yahshua came to earth he arrived as a Judean baby, and he grew like we do, and spoke the language

6. In the words of the famous physicist Feynman describing the great *ifs* of probability that describe light and matter in a laboratory, "The theory of quantum electrodynamics describes Nature as absurd from the point of view of common sense. And it agrees fully with experiment." Feynman, *QED*, 10.

of the people, and he bled like a man. We love him so much; we know we do. His empathy for us resonates with us, and a Christian reciprocates by empathizing with his suffering as well.

When Yahshua purchased our debt, he proved he loves us. And Yahweh proved that he *requires* justice (consequence for sin) and Yahweh will not even allow his sons and daughters (those chosen sinners whom Yahshua purchased) to escape justice (consequence). Yahshua is our priest and king, the chosen one we have always been seeking, and Yahweh is that perfect judge whose reliability is calculable. If Yahweh did not require justice, would he be worthy to sit upon his throne as the judge of angels and men? If one of our judges here on earth allowed his children to escape justice (to suffer no consequence for crime), would we respect their authority? Forgiveness is a wonderful blessing, but not at the expense of justice—the two must exist in harmony. Yahshua is a living example for how justice and forgiveness exist in harmony in our world that Yahweh judges.

The harmony is accomplished through זבנא ("exchange"). Blood has a fundamental purpose in life and atonement, as we read for example in Leviticus 17:11.

I imagine the angels in heaven calculating the fullness of each person's experience, a system of just weights and measures. Per John 10:18, Yahshua was given power on earth to lay down his life and to take it back up again. He is more alive than anyone on earth, which meets the prophecies of Ezekiel in Hebrew (I think it helps to read לא ימות in Ezekiel 18:21 as "not losing life," which is a perspective on death that the context of Ezekiel seems to provide).

Many who seek have found that the word is represented in our world through an aleph bet that concludes awesomely with the letter tav in such a way that tav represents a cross. In the original Hebrew script, tav was indeed a cross-shape. I think that cross shape is just the letter we see on the surface. Beyond the surface, the tav is multi-dimensional, just as all the 22-letters are multidimensional energetic waveforms that flow throughout the torah/torus of our world. Some very fascinating research has shown the heart's electromagnetic pulsing represents exactly the 22-Hebrew letters.[7]

Regardless, when you hear the Messiah's story, you feel sorrow and joy in the correct proportions as you are in harmony. You know if the

7. This was demonstrated by researchers like Stan Tenen, who sued Dan Winter, which had the effect of focusing or limiting the application of this knowledge. I consider it likely the research is correct that the aleph bet is written on the human heart. See also Meru's focus on the aleph bet as hand gestures. Kirsch, "Hebrew Alphabet," para. 5.

Messiah resonates with you, as he admonishes those he loves. Revelation 3:19. Yahshua is ultimately the hero in this story, our world, a world that was born from the Word.

From a physics perspective, I think the cross represents two lines that come together, two waves travelling perpendicularly—do the lines collapse? No. Will something emerge or be released (like light) from their interaction? Yes.

The meaning of synchronicities is that all things are interconnected and weighed and reconciled by the law/ratios that provide meaning to the concept of exchange itself. The tav/cross is about exchange because at the conclusion of the story, there is a remainder, a matter that is reconciled. God sets the laws/ratios, which to our perspective as humans appear within a base ten number system that instructs us to the reality of the numbers 0 and 1 through 9 (not incidentally the Hebrew word God, 19 = אלהא in gematria). If you follow the law/ratio given by God then at best you live like a fractal of the Messiah. If you transgress the law/ratio, you collapse like a wave collapses, in measure. It is the character of the Father to balance equations—in this manner, the cross also represents a scale.

On the surface, letters are like forms that instruct about the spacetime mechanics of our world, and the meaning of our story. Many people go their entire lives constantly amazed, and correctly so, by the interconnectedness of worldly events and the synchronicities in their lives. The synchronicities provide relative meaning to the mind. We seem to have the power to point our minds in the direction we choose, to focus. This teaches us about the law of perspective.

Through letters we are learning about patterns and concepts. Earth is like a classroom in that way. For example, the Aramaic spelling of Caesar is *qsr*, which reminds of a quasar, a luminous body powered by a black hole at its center, such as the M51 galaxy with a cross-shaped image at the center. There is likely more to the cross/tav and this battle of Yahshua versus Caesar than meets the naked eye.

We define reality by *source*. I look in the mirror for example and I consider the reflection less real because I know it is not the source, but only an expression/effect of the source's actions. We humans have trained our minds to gravitate toward the source of what makes us feel/experience. This helps explain why we gravitate toward the living Word more than ink on paper. We know aleph as a kind of source, and tav represents the conclusion of a cycle—we found meaning in the journey, we seek desperately to find in ת a door to source א. This is the אתא ("sign") given to us. Text is merely a tool.

We learn about the lamb through examples in the story: Abel's sacrifice (a lamb) leads to Cain's second sacrifice (Abel). In Egypt the Passover blood over the door (to חורא represented "freedom" for a price). In Jerusalem the selling of lambs in the temple became so commercial that even the gentle lamb who would not break a reed actually tipped over the tables and picked up a חבלא (whip); note the wordplay חובלא = debt/burden/cargo.

Abel sacrificed a lamb, and we learn later that the lamb was a symbol of his body. Cain was a different kind of priest—his תנינא ("dragon/second") sacrifice was his brother! Recall Matthew 23:35, where Yahshua implies that Cain was the original Pharisee *and scribe*, "In like manner, that will come upon you, any of it, the blood of the righteous that has been shed upon the earth, from the blood of Abel the righteous and up to the blood of Zachariah the son of Brakhya, he whom you killed between the temple and the altar. Truly I say to you that will come any of these upon this שרביתא."

This word שרביתא is normally translated "generation," but I think here it refers to a line of power.[8] A line of power was given to mankind—it is our ability to manipulate the word of creation (i.e., Adam was given the power to name things; Moses was given the power to write laws). Cain used his power to slaughter Abel. Some Pharaohs used their power to slaughter the Israelites. Whenever a man sets a command/action upon his brother that he does not want himself, he slaughters/divides. These are patterns in the classroom of earth. We are learning how to add and make things, and to subtract and destroy things.

The nature of the flesh is an insatiable desire for the consumption of energy and information; we are perpetually unsatisfied creatures, as the simple example of Eve's apple revealed—when we wield our appetites properly it makes us interesting (i.e., we make things like cakes, architecture, clothes, machines, etc.), but when we wield our appetites outside the natural law/ratios, then the results are destructive from within.

Why does Matthew 23:35 suggest that Cain was a scribe (or that his actions are attributed to a scribe)? We have no record of Cain writing anything, or that by killing Abel he was following the written directions of some scribe. I think the answer is that Cain is a scribe because every man is a scribe by the very nature of speaking or even moving a finger—*everything* in our world (even the breath in our lungs) is made from the letters

8. Glaser, "Tribal Conflict or Harmony," 1.

of life—the world is made of letters. Every time we move a hand (with or without a pen), we are writing something because we are literally made of letters (energetic waveforms). My heart is made from word, your hand is made from word, and the output of word is word. Word upon word upon word = story = number upon number upon number. The equation/story is balanced by God.

So, we are all scribes and priests. And the messiah is the head priest upon the cross/exchange who reconciles our equation (division of words) in the following manner: the two lines of the cross intersect and partially cancel one another in their זבנא ("exchange," "time"), but a זרעא ("seed") portion of each line braids the other and survives in the זבנא ("exchange," "time"). The result is that a new world is born in the Messiah.

Why a lamb? The word for "lamb" in Hebrew/Aramaic is אמרא, which is the same word for "speak." Per Genesis chapter 1, our world was *spoken* into existence by God. So, the division of a lamb represents the division of a spoken word. Also, a clean lamb is white, a color that represents the full visible electromagnetic spectrum. The division/sacrifice of a white lamb represents the division of colors to our minds, which helps us learn to distinguish things in this world and appreciate reality here. Per John 1:4–5, distinguishing light and dark is essential to the human experience. Distinguishing all colors is good.

But why did this white lamb, a fragile living being, need to be sacrificed rather than enjoy a nice retirement on pasture with fresh vegetables? The nature of storytelling and life is that beauty is found in the saga and triumph of vulnerable things. So, if you are a parent telling your son a story, and you need something white, fragile, small, living, and that represents word and more in Hebrew, then a lamb is the natural choice.

And the story of this world has a happy ending, because the lamb/son/seed survives. Division does not destroy God's son. The ratio triumphs, the son lives in the Father, the seed survives, the torah renews.

There are many ways to learn a lesson. There are many ways to tell your son a story about himself.

On my ranch I have had to slaughter a lamb and clean the balance. The immediate feeling you experience is sadness. You also feel thankfulness for clean air and water that helps you accomplish the job that is necessary. As the shepherd, you hope the lamb's death was not in vain. However, if the lamb's death was due to your own error (i.e., your dog/appetite escapes your fence/control), you feel great shame. In my experience, slaughtering a lamb is not something any normal person desires to do unless it is necessary. If you had to slaughter lambs regularly (like a modern butcher or ancient priests in the temple), this would take an immense toll on your psychology and make you unbalanced (today, psychological studies on butchers help support this conclusion).[9] But learning about the experience/concept itself is useful to a balanced psychology.

As we exercise free will in a world of written laws, the tension is palpable.

This book asks whether reality exists because of the energy-creating tension between paradoxes. It is paradoxical that we find aleph in tav, life atoned by death, and God is so great that he is also the Nothing. We observe paradoxes are creative forces, and we rightly expect their prevalence. Living the tension between free will and law in our own lives, we learn about ourselves. Yet can we even imagine the tension at God's level of paradoxical interaction? What is God's freedom in Randomness juxtaposed with his will in everything else? Rhetorical. We need God to teach us.

9. See e.g., Slade and Alleyne, "Psychological Impact of Slaughterhouse Employment," 438.

Chapter 6

Golden Ratio Paradoxes

SUMMARY

RATIO IS HOLY. THE reason "one" is "one" is because of its *ratio* to zero, meaning "one" is precisely "one" *from* zero. If you remove ratio from something, it becomes so abstract as to become unknowable. Music is ratio. Even letters and words are arranged in ratios, yet the scales are not set by physical laws per se, but by consensus of men. As we discover that men are fundamentally built upon the golden ratio, we discover our words are also ratio. The golden ratio is that special number found throughout life in the universe. It is everywhere in the atomic realm, your body, DNA, the earth, and the stars. It is key to understanding life. It also provides perhaps the best analogy to our relationship with God: the whole is known by knowing the relationship of the larger part to the smaller part.

Relevance for your life . . .	Imagine for your life . . .
You love healthful natural systems. Therefore, you love the golden ratio.	Your son is to you as you are to your family. Every relationship you can get into golden ratio is improved, because God chose the golden ratio to order the universe.

> **Logical Hypothesis**
>
> a/b = (a+b)/a
>
> a = primary
>
> b = secondary
>
> a and b = whole

INTRODUCTION

The golden ratio (phi) is simultaneously simple and complex. Phi occurs in a system whenever the ratio of the smaller part to the larger is the same as the larger part to the whole.

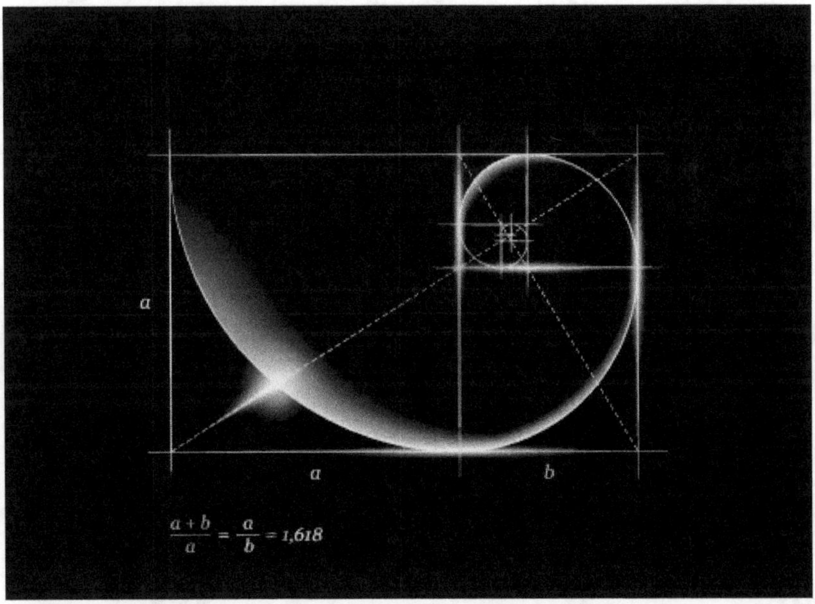

It is universally recognized that phi is beautiful and extremely powerful. For example, the most attractive human faces are in phi proportion,[1] and likewise for nature, architecture, art, etc. Many books have been written on the matter. Does all this beauty and power suggest phi is in our

1. Meisner, "Beauty/Face," para. 1.

source code? Yes, our DNA is an excellent example of phi,[2] and DNA is an ordered and fractal braid.[3]

So, neurologists have shown that our minds inherently recognize the beautiful proportion of phi. We gravitate toward it. Biologists have shown its presence as a ratio throughout life on earth (e.g., pinecone, sunflower, your eyeball, your hand, your entire body if you are proportioned well). Even cosmologists are increasingly discovering the ratio throughout stars in the universe.[4]

The bottom line is that phi evidences stability and repeatability of a living system. Phi is also so powerful it excludes/destroys that which is non-phi.

In this chapter we will also consider the wonder that phi withstands charge compression (i.e., survives death). This is important theologically. Practically too, if you want to survive an extreme event or thrive in conditions conducive to life, you will be glad to have .618 in your navigable ratios.

It is no accident that in the ancient language of Aramaic spoken by the apostles, the oldest recorded text of Matthew 7:12 (the golden rule) has a gematria numerical value of 618 exactly, which is the number representing this golden ratio (phi = .618), also called the golden number or divine proportion.[5] The golden rule is probably the most universally accepted of all Christian philosophies. In ethics, the golden rule is an expression of the golden ratio. Per Matthew 7:2, whatever you measure is measured

2. Perez, "DNA Tuned by Golden Ratio," 228. See also, Merrick, "Harmonically Guided Evolution," 2–5.

3. See e.g., Blank and Goodman, "DNA Is a Fractal Antenna," 409. When you listen to a Mozart symphony, you know inherently that the silent spaces are just as important as the notes being played. We call this harmony—so too with DNA. The term "junk DNA" is a misnomer because the so-called empty spaces preserve the integrity of the structure in a ratio that is fractal. And as highlighted in this chapter, being a fractal that produces a 1.618 ratio in its component parts is important for charge compression and expansion in the measurable universe.

4. See Val Baker, "Golden Ratio Stars," para. 4. Meisner, "Cosmology," para. 1. Boeyens and Thackeray, "Number Theory," 1–2. Pletser, "Prevalence of Fibonacci Numbers," 25. Tattersall, "The Hum," 194.

5. In addition to words, the ancient text of the gospel can be read as a string of numbers (22=ת; 2=ב; 1=א). This is called "gematria". Using gematria, you can calculate the numerical value of any passage of the gospel, and the golden rule in the Peshitta calculates to 618, the golden number. This means in the ancient Peshitta text of Matthew 7:12, where Yahshua gives the golden rule ("All that you desire that men should do to you, so also do to them, for this is the law and the prophets"), his words add up to 618 exactly.

back to you. By participating in a fractal relationship, you validate your own existence.

In mathematics, there are countless ways to describe and utilize a golden ratio perspective.[6] Everyone loves phi, whether they know it or not! It is absolutely fundamental to life and philosophies about life. By using the golden ratio in inventions and engineering applications, we maximize efficiency and increase our chances of survival through sustainability. That is important to daily life. How better to validate our existence than by surviving from generation to generation?

In this book I place a special emphasis on the ubiquitousness of phi as *the* primary fractal pattern of life. However, it should be noted that conventional science tends to categorize the golden ratio with less importance, almost as if it were just another fractal. Most cosmologists for example who study astronomy routinely do *not* look for phi in the systems they study, so they do not report phi's presence or influence; yet the astronomers who do look (as cited in this book) find. So I expect phi will continue to grow as it does best.

PHI SOLVES A PARADOX

In chapter 1 we reviewed a *paradox of order* referring to the golden ratio. We learned this special number (1/1.618033) reveals order in oneness (the whole is known by knowing the relationship of the larger part to the smaller part). Because you cannot count something twice, the paradox is that perception is simultaneously real (tangible, countable) and unreal (holographic, hypothetical).

In other words, the paradox is that you can observe the whole (Universe) or the parts (i.e., people), but you cannot observe both together simultaneously. Either (1) you focus on the parts individually, in which case the whole is just *an abstraction or concept*, or (2) you focus on the whole as a network, in which case the parts are just *a perspective* on the whole.

In option two—if the parts are just a perspective on the whole, then a simple conclusion follows: as a part, I am not real—I do not even exist separate from the whole. My entire life and psychology is just a *perspective*, perhaps even a glare in a mirror that can be eliminated at any moment by a shifting position.

6. Bogomolny, "Golden Ratio in Geometry."

So that is a paradox of perception—and fortunately, it is resolved by the golden ratio. Consider deeply the meaning of phi. This golden number occurs in a system whenever *the ratio of the smaller part to the larger is the same as the larger part to the whole.* As a fractal, the golden ratio allows both the parts and the whole to co-exist within one another without collapsing (that is, without destroying each other). That is why it is sustainable, and therefore why we find it represented in life throughout the universe.

We can solve the paradox with phi. If I observe the whole then I see a large fractal shape (such as a Fibonacci spiral), but the ratio that comes from the parts *cannot be removed without destroying the whole.* And if I observe the parts individually, then I see the whole because the whole is embedded in the parts; and so likewise the whole cannot be removed without destroying the parts. Phi is the embodiment of coexistence—living within one another.

Thus, to the extent we are in harmony with the golden ratio, we are living proof that we really do exist and that our perspectives are valid.

GOLDEN RATIO EXPLANATION FOR SURVIVAL OF THE FITTEST

On this planet we observe survival of the fittest, and the "phittest" is phi. For example, a predator that kills too much is not the fittest (though it may be the most ferocious) because it eradicates its own food supply. Phi respects balance. The fittest predator is the one who naturally and symbiotically consumes and produces its share of the earth. Not too timid and not too aggressive. As discussed earlier, consumption and production are ultimately the same phenomenon. A strength in one context can be a weakness in another.

The reason our reality manifests as survival of the fittest, and trial and error, is because (per QED for example) light and matter travel *all possible pathways.*[7]

So, our world can appear imperfect in the sense that we need to navigate among many of these undesirable possibilities; we experience

7. In other words, mathematically accurate results are derived from measuring all possible paths of light and matter. So, if the math matches reality, then reality surprises us with complete quantum interconnection. Why would reality have unexpected qualities? Laughter is also a form of surprise—why is wholesome laughter so fun and healing? God is our Father who edifies us.

the weakest in order to work toward the fittest. And we navigate among "errors" in order to create "success," yet paradoxically the successes are constructed from so-called errors. This begs the question—in a world of relative contributions amidst complete quantum entanglement, is anything ultimately erroneous?

And in the aggregate/macro of this process, we observe the *most likely* outcomes. Weakness and vulnerability are most likely outcomes until the fittest triumphs in an environment. And once it triumphs, it sustains itself (phi sustains itself). Also "error" is a most likely outcome until "success" triumphs. But the result is not static, as "success" in one moment might be "error" the next. And success is constantly surrounded by error. It is ratio (the abstract) that transcends the physical success and error results.

The fittest and success are manifested through the golden ratio because phi grows sustainable life into the environment in such a way that excludes other outcomes in its defined space. It is both a simple and complex wonderment that with so many environmental factors in play (affecting growth), that phi would be such a pervasive success.

And, as a fractal, the golden ratio allows both the parts and the whole to co-exist within one another without collapsing on each other (again, without destroying each other). That is why it is sustainable, and therefore why we find phi represented in life throughout the universe. John 8:3, "Yahshua answered and said to them, 'Even if I testify about myself, my testimony is true, because I know from where I have come and to where I am going.'" Likewise with phi, it is certain/calculable where the energy has come from and where it is going.

The reason the golden ratio is so successful is because it is mutually beneficial—the two fundamental parts of any given interaction (one large part and one small part in 1:1.618 ratio) focus exponential possibilities into the service of one whole, while this whole receives their service and gives them greater meaning and conceptual purpose. Beautifully, it is a metaphor for our relationship with God.

So, what are the paradoxes in play? Is the whole both greater and lesser than the sum of the parts? Do the parts unify to become the whole, or does the whole entropy into the parts? If entropy and unification are the same phenomenon, such that the outcome is ultimately the same, what is truly the fittest? These questions will help us discover a new paradox, where the golden ratio reflects on itself and therefore becomes a new and valuable thing: memory. But is the memory real? Is it inherently valuable?

There was a famous philosopher named Sextus Empiricus who discussed a paradox about probability, as Sorensen explains:

> "Sextus thinks it is dogmatic to affirm that probability is the guide of life. To change your mind on the basis of probabilities, you need to assign some probabilities prior to any inquiry. These prior probabilities are assigned without any reason. But then one is assigning some propositions higher status than others without any basis. This favoritism is dogmatic. Sextus casts himself as the open-minded inquirer who refuses to acknowledge that any belief is more probable than any other. Since he does not want to commit himself to any proposition, he does not want to assert that we lack knowledge. For all we know, we know as much as we seem to know."[8]

A problem with Sextus' reasoning is he uses the same multiverse logic (everything is equally probable) that we exposed as false in chapter 4. As professor Karpenko correctly observed, if everything is equally probable, then probability becomes meaningless and does not exist. We clearly observe probability exists in physical reality and it correctly predicts the future.

CHARGE COMPRESSION

Phi is not infinite, but rather it is repeating. A tree is a good example of phi—it has repeating qualities throughout (roots, branches), and it produces seed that allows repetition after death. A tree is not infinite; it is limited in both space and time.

Phi is fractal, meaning you can look at any part of phi to discern the whole.[9] Sometimes it surprises too, such as an intricate flower emerging from an otherwise simple looking plant. If something is not fractal, then when it compresses after enough iterations it eventually collapses on itself and is "destroyed" in the sense that the original shape no longer exists. We say though that energy is not destroyed, just transferred. So, the "transferred" energy is given to the forms that *do* thrive. Matthew 25:14–30 (parable of the talents). It is the fractal forms that thrive. Studying and modelling fractal forms is a way to study life optimistically.

This relates to *charge compression*, which is a concise expression for the measurable electromagnetic rotating energy (charge) that collectively

8. Sorensen, *Brief History of Paradox*, 153.

9. Phi is not the only fractal. In some fractals the whole cannot be discerned from isolated parts but rather the pattern repetition is at different scales.

shrinks into a smaller volume from its present size. When the properties of that object all shrink together at the same ratio it is a kind of fractal charge compression.[10] When they all expand at the same ratio it is a kind of fractal charge expansion. Surviving compression and expansion is a fundamental trait of living things. It is how seeds work so that plants survive death year after year (rotation).

One way that I would describe an infinitely dense singularity is that the *probability* of its density appears infinite because no measurements/ focus have been placed upon it. (Without an observer, a wave does not reach its destination/purpose so it retains probability in order to function as a wave). To the extent that infinities appear on paper in our mathematics, they can be signs that either the wrong tool has been used for the job, or not enough tools have laid focus upon the *probability spectrum* in question.

We tend to believe there is something missing with the theory that the universe came from a state of Nothingness. In other words, the problem is "where does probability come from?" It seems too convenient to say that

10. A fractal charge refers to measurable electromagnetic rotating energy (charge) flowing throughout an observable fractal shape in such a way that it appears to (or does) form that fractal shape. Fractals are fun and have been essential in shaping science. See e.g., Lisle, "Mandelbrot Set."

paradoxes seed themselves as they strive for infinitely more focus within one another, yet never obtaining it—a perpetual state of imbalance in the very nature of form and movement.[11] Even if we do away with the word "infinite" by replacing it with "improbable," we still have the same question just highlighted—where does probability come from?

Scientific Speculation About a Plasma of Probability

If there really is a plasma of probability at the foundation of physics (as theorized in chapters 1 and 2), how could it be tested or observed? The answer may be fractality. To move beyond the uncertainty principle, physicists might someday be able to experiment with tools of a smaller size than the photon. This is a physical principle for avoiding wave interference, and another reason why focus is so important in QED. To compress a tool/wave into the smallest quantum realm, physicists would need a kind of crank that expands and shrinks fractal structures up and down and across in classical 3D space (and by default into the quantum realm) with perfect uniformity—predictable fractal models would therefore test the assumption that fractality continues into the smallest quantum realms. The key would be that fractals create predictable patterns of interference. That predictability of wave interference is essential for modelling something you cannot see (below the Planck scale).

In other words, for an observer like a physicist working at the classical scale, he needs a model that calculates exactly where and when a balance is disturbed (at the classical level, and confirmed repeatedly up to the Planck boundary). Therefore, by modelling disturbances of fractals in this manner, I assume physicists can make deductions about fractal energy operating at the smallest quantum scales—they measure how it affects balance at our more readily observable scales.

In a primordial hypothetical Randomness, in a plasma of probability, one can theorize a hypothetical wave particle: possibilitrons. They can be on, off, both, neither. They have angle, spin, direction, speed, rotation, and more. In golden ratio, the possibilitrons become a self-aware network. For example, from God and/or Randomness, possibilitrons emerge as phi, leaving an imprint (or writing) because the networking among possibilitrons

11. In addition to creating perpetual imbalance (hence energy, existence) paradoxically a paradox also creates structure (hence existence). Example: two planks resting on each other like this / \

itself is an entity (vibrating wave particles at the same resonance creates a *new* effect). The imprints eventually create a form, which itself is probability. I am speculating.

Consider Boni, "There is simply no way to predict with 100% accuracy which wave-function potentialities will be physically actualized in spacetime, and which will collapse away into nothingness."[12] This suggests the wave potentiality is 0 to 1. Upon focus every possibility becomes 0 except 1. Boni then continues, "The reason being, that quantum event outcomes simply have no physical determining causes, such that there is no antecedent causal chain to flow either backward in time or forward in time as was the case in Newtonian mechanics. The problem, then, is quantum a-causality, and it underwrites every single particle in the universe . . . What *really* determines a quantum event? [T]he answer is still 'nothing'. Regardless of how we may feel about it, quantum events are a-causal—at least, according to the physicalistic rules of naturalistic science."[13]

Further Speculation and a Proposed Experiment

I speculate that the absence of fractality will prove why our human bodies decay/age over time. Imagine a vibrating string (wave) inside a box (3D space) and its field (mass) and form (energy measured geometrically) appear as a tangible shape (like the letter "c" for example) to the perspective of an observer that collects information from the surface of the box. In one view, that box's energetic values (e.g., mass) would be observable/measurable to us with countless other boxes in pixelated form. Cellular aging could be measured then as copy errors summed from each pixel due to probability variations (examine the wave pattern to detect variation).

In this viewpoint, our original DNA wave strand code is a source code, and its repetition in unmodified wave format promotes a uniform cascading phi ratio. For length of life, you need like-form inputs promoting your fractality (so you consume similar phi ratio objects like organic food) to promote uniform cascading. Whereas interference with the wave ratios (through GMO-corn for example)[14] eventually lead to mutation and death after enough iterations are allowed to multiply the effects.

12. Boni, *Quantum Christian Realism*, 16.

13. Boni, *Quantum Christian Realism*, 16, 18.

14. Consider recent studies of mice that consume GMO food—after several generations of DNA repetition, the results are not pretty (cancerous tumors, infertility). Smith,

Getting back to the point about pixels (which incidentally our computers are well-suited to model)—focus, use, and juxtaposition are words we use to describe how strings materialize on the surface of pixels in every iteration of change/movement. Just as general relativity emphasizes the importance of directional change in space, so too we could predict that directional change is necessary at the quantum level to force a probability-changing event on the surface of the pixel. On the macro-level not all things are fractal, but on the micro-level perhaps there is a point where all is fractal. That is where you might expect to find a source code written like phi. And if we learned that such a source code exists, we might start explaining consciousness as the way that wave harmony and wave variation are dependent on what we focus upon to fulfill the source code.

Focus forces choice/change because focus requires extraction of information from the wave, and the extraction/interference changes the wave because the observer is made of waves too. Extraction is a type of reflection that results automatically in a wave environment because each part of the whole automatically measures itself against the whole—recall the scale analogy in chapter 2, which is another example of ratio).

I further speculate that the only way to avoid extraction of information (collapsing the wave; changing the probability) is with mini-instruments (i.e., tools smaller than the photon) or extra-dimensionality (i.e., reaching into a sealed 3D box from the fourth dimension) or oneness harmony (a spiritual concept).

In the quantum realm, I would like to know whether motion is sufficient to create energy or whether energy is an illusion created by shifting probabilities. In one view, motion is needed to create energy. In another, motion comes from perceived focus on different aspects of unified energy within a complete torus (or multiple tori in balance reflecting on themselves) that is the entire universe. In this latter view, perceptions change based on probabilities. The probabilities come from some code/law that is either ordered or random. A random code generator would be a simple expression, and in time the code would spark structures of reflection that could sustainably reach fractality and create order and intelligence, and even a kingdom of heaven.

If scientists recreated such an environment (i.e., order from randomness), it would be *additional* physical evidence of God. And further the experiment would require faith, because the data received from beyond (smaller

---

"Science Guide."

than) the Planck length would not be *directly* observed but rather only inferential, assuming the scientists are continuing to work in three dimensions.

GOSPEL APPLICATION

How does phi relate to the gospel message? I will offer a theory or analogy that is definitely *not* intended as dogma—if you are a smaller iteration of phi energy vibrating ("phi-brating") and you are sent on a mission from your harmonious larger iteration of phi energy, then you will *accurately represent* the larger in the way that you (the smaller) interact with beings even smaller than you. In doing so, you (Yahshua) will reveal to those beings (lost sheep of Israel) how they can begin to know the larger (God) and the whole (the universe) by first appreciating the lesser (Yahshua's example). In this theory of interpretation, phi is like a narrow path that is beautiful and creates optimally efficient life and structures in the observable universe, whereas the path outside of phi is wide and that path is scattered and inefficient and does not lead to life. So, phi is a key *part* of the character of the kingdom of heaven, just as certain Fourier transforms (to choose just one example) would be a part of that mathematical character.

Yahshua said in Matthew 7:2 that whatever we measure will be measured back to us. We can be literal with that—we can measure phi and know it will be measured back to us, because it is found throughout life in the known universe. To seek Yahshua with honest hearts it can help to recognize his character as a physician healing the sick. Phi heals the sick. To seek the Father fully it can help to recognize his character as a physicist completing our equations. Phi completes equations.

Perhaps if we dare add even a single aleph to the gift of the gospel (as many later texts do with Matthew 7:12), we run the risk of distorting our appreciation of Yahshua's energy harmony with the Father. And in a world where the golden ratio *evidences* life, our transgression from the literal word puts us further out of harmony with Yahshua (the lesser), and therefore out of harmony with his Father (the greater), and therefore out of harmony with the whole (universe), to the extent we seek life. Again, this is offered as an illustrative and imperfect example, not a dogma. This example helps explain how the Father and Son are one, yet the Father is still greater.

THE SMALL BECOMES LARGE

The paradox: if you want to grow large, you must be small.[15] Following the theme of this book: the two truths must be the same thing. Example: even one tiny sperm can become so large as to create a human who creates a population of humans who seed the galaxy and who covenant with God Almighty. One little sperm.

Because phi is so intimately connected with life and reproduction, it can be a bit awkward to discuss, but here we are. There is a mystery within golden ratio math that points to the female womb as a holy grail. Scientists observe the golden ratio in the uterus.[16] The human seed (sperm) is extremely small, smaller even than the mustard seed. The initial smallness of the kingdom of heaven is emphasized repeatedly in Matthew, and if you read the gospel carefully you can see the connection to seed and the golden ratio:

- "The kingdom of heaven is like a mustard seed that a man took and sowed in his field. And it is small among any of the small seeds but when it has grown, great it is among any of the small herbs and it becomes a tree so that the birds of heaven come and nest in its branches." Matthew 13:31–32
- "The kingdom of heaven is like leaven that a woman took and buried in three measures of flour until any of it was raised." Matthew 13:33
- "The kingdom of heaven is like a treasure that was hidden in a field that a man finds and he hides it. And from his joy he goes and sells anything he owns and buys that field." Matthew 13:44
- "The kingdom of heaven is like a merchantman who sought good pearls. And when he found one pearl of precious likeness, he went and sold anything that he owned and bought it." Matthew 13:45–46

In Matthew 13:44 the kingdom of heaven is likened to a treasure buried in a field. In the Aramaic language, the קריתא ("field" and "beam") comes from גלא ("reveal" and "straw" and "wave"), which is illustrated in Matthew 7:5. Imagine this "treasure" was like a golden ratio wave hidden in a "field" (the many waves of this world). In Aramaic, the word סימן (pronounced *semen*) means "treasure," and humorously a grain of wheat covered by chaff does indeed look like a sperm.

15. Recall Mark 10:43 ("he who desires in you to be great will be to you a servant").
16. Bellos, "Golden Ratio in Uterus," para. 13.

In a rebirth, the heavenly womb is for those who are small, not large. Heaven is for those willing to conform themselves to the pattern (the seed) of God, and then when their appointed time comes, the flesh passes away but the seed given from God is reborn. John 3:5–8 emphasizes the importance of being born again from Spirit.

We learn from the example of children as they are close to the truth/source in their own manner. Learn from the example of the humble, just as Yahshua explains in his sermon on the mount in Matthew 5:5. The womb is peaceful: blessed are the peacemakers. Matthew 5:9. The heavenly womb follows the law and delivers the baby according to the code of the seed, the ratios. On earth, we learn from the example of DNA following the golden ratio.

Conforming to the code/law of God is essential for life—it is why instruction and admonition are literally *essential* to understanding the kingdom of heaven in God, so you can point your soul toward God. Deuteronomy chapter 28 makes it clear this is a matter of life and death.

Learn from the wordplay in Matthew 13:45–46 with regard to the kingdom of heaven as a pearl, where pearls are admonitions.[17] Be like a child and follow the rules. A child in the womb is the best example of all—he conforms to the shape of the womb and grows in the golden ratio—a son of man follows *exactly* the code/law given by his father and mother. He is born clean.

To understand the concept of rebirth from the mother aspect of God's creation (see e.g., Proverbs 8:22), you are given the opportunity to connect earthly concepts (sexual desire, church/assembly, this world) to heavenly concepts (spiritual heart, kingdom of God, new world birthed from heavenly spirit). Nine months after sexual intercourse with a woman on earth, there is a new beginning—a child is born *clean*, a son of man is born again in flesh. I think this is why after eating bread (a Hebrew euphemism for sex), Yahshua explains to Simon Peter he needs to "wash his feet" (another Hebrew euphemism for sex) in order to become clean.[18] Simon Peter and

17. Glaser, "Wordplays and Segways," para. 2.

18. Hebrew euphemisms are often funny. See the many examples provided in Rendsburg, "Word Play in Biblical Hebrew." For example, eating bread is a euphemism for sex. See e.g., Genesis 39:6–9, Exodus 2:20–21, Proverbs 6:26, Proverbs 30:20. Feet are a metaphor for a sexual organ. See e.g., Ruth 3:4, Isaiah 6:2, 7:20, 2 Samuel 11:8–11. Knock is a metaphor for an invitation to sex. See e.g., Song of Songs 5:2. Hand יד is a metaphor for the male sexual organ. See e.g., Isaiah 57:8. זכר is a metaphor for sex. See e.g., Ezekiel 16:17, Isaiah 23:16, Isaiah 57:8. Interestingly, John the Baptist is the son of זכריא. If crossing the Jordan is like washing your feet in the river, how much more a baptism?

others may have even chuckled at the back-to-back euphemisms. But there is a serious spiritual aspect of rebirth that creates fullness. The fullness includes a mystery—the spiritual mathematics of rebirth create sustainable fractal revolutions. Those chosen to hear Yahshua are given to look beyond this earth in their own unique ways, to spiritual rebirth. The flesh is a sign. Sometimes we are drawn to it for constructive reasons, and sometimes destructive. Sometimes we are repelled from flesh for constructive reasons, and sometimes our repulsion destroys us.

Consider this example—you experience that you have a healthy and loving relationship with your wife if you want to share sexuality with her in a sustainable way: marital fidelity. Wholesome sharing is at the essence of the healthy desire for sex, because that is what love is made of (in simple terms). It has been said that if people could see the energetic waves that they emit, they would be very careful what they *think* while consummating a sexual action. Notice how the word for "sharing" שתותפון in Matthew 1:18 is utilized to identify the *marriage* of Joseph and Mary after Mary receives the Holy Spirit.

Also, the Hebrew word הבר ("sharing"), in its component parts בר is a reference to son or daughter, and the ה prefix emphasizes life. Love between a man and woman is supposed to be about creating a child, wholesome.

TORUS AND LIGHT

Imagine darkness was synonymous with randomness (i.e., because you can learn nothing from randomness alone, you are literally in the dark). And imagine light was synonymous with order (i.e., light produces life in phi ratio). Are dark and light two competing truths that are actually the same thing? In Genesis 1:4, we read that God *separated* dark from light. This implies that dark and light were originally unified in some way before God separated them. Job 12:22, "He reveals deep things from darkness and brings deep darkness to the light." Daniel 2:22, "He reveals deep and hidden things. He knows what is in darkness, and light dwells with him."

And consider the research of a man Nikola Tesla admired, Walter Russell, who posited that all is light (even darkness is just elongated light waves). This is similar to cold being the relative *absence* of heat.

The separation of light and dark in Genesis 1:4 could have been accomplished by setting ratio, setting law. Was phi utilized to accomplish the separation, and if so, will phi be used in the future to unify? The Russellian view is that the universe is the illuminated outward manifestation of the mind of God. Or in other words, the electric (+, north) magnetic (-, south) universe exists within God and is secondary to God.

Recall the Bible's first word בראשית ("In the beginning"), and my alternate translation "It was in his head". Imagine explaining the origin of the universe as a hypothetical torus (God's light) emerging from conceptual Randomness (dark). The torus harnesses the conceptual Randomness by consuming the energy in itself, to expand itself and multiply. Consumption and production are paradoxically the same thing.

I envision that our material world exists like the face (outward manifestation) of a torus of existence, and the quantum world is the small core of that torus. Or alternatively, the two worlds are each their own torus, existing holographically as mind. You can imagine a torus revolving inside another torus and vice-versa, like an Escher drawing in constant paradox.

In John 16:28, Yahshua describes how he נפקת ("departed") from the Father to come to this world, and now he is leaving this world to go back to his Father. נפקת is a play on פקד ("commandment") because תוב שבק ("leave again") in this verse can also be translated as two commandments, "repent, forgive."

And to the extent Yahshua is light, then John 16:28 is describing light mechanics—light from the Father reflecting off earth, back toward the Father, but some בדר ("scatters") in John 16:32 (sheep scatter when the shepherd is seized). Even the letters קד here could be a clue for "quantum electrodynamics" (the study of light in quantum physics). Like Russell, my inclination is not to limit the possible connections between concepts, but to consider the exponentiality of interconnection.

Imagine John 16:28 is like a physical example, where light leaves the center of the torus (symbolizing the Father) and goes out into the world (the face of the torus), and then it returns back to the Father (center) because it is fractal and obeys the פקד command of fractality.

It can be interesting where the Aramaic hints at physics in the gospel. But it is important to remember that analogies are just analogies—obviously God is not simply a torus or some physical thing. Rather, understanding torus math helps a person learn *about* God. The torus is just a perspective for learning about God's greatness. For example, everything outside the center of a torus bends down (alternate translation "worships") toward the center of the torus. Is that a mathematical reason why men סגד ("worship," "bend toward," "honor") God. Or if they do not, then are they still being fractal (law abiding)?[19]

Even though .99 and .01 are far apart on one axis, they are similar on another. They are both extremes, close to source. They are as far as east and west, yet as close as the center of a torus. So, the verse "the last will be first, and the first last" (Matthew 20:16) can also be quite practical from a math perspective, in journeying toward the center of a torus, or the flipping of its rotation.

Knowing God *is* life, per John 17:3. Light grows life and can help us know God. It is weightless, without cargo, and yet it carries information able to create more than cargo. Light is revealing in waves, and light is faithful.

19. In Hebrew, the sound "math" מתת means "gift." But "math" מת can also mean "dead." Beware self-proclaimed rulers in this world, such as the ancient Pharaoh and the modern Pharaoh that dazzle men with mysteries and require their servants to bow down before them. Glaser, "Wordplays and Segways," #3, paras. 3–5.

SQUARING THE CIRCLE

Mathematicians have said it is impossible to square a circle. Does it matter that in a world of dots, a square and a circle are actually the same thing? The square is constructed of dots in straight lines. If you spin even one line (or the four lines making a square) on a central axis in every possibility it appears as a sphere. Thus, the cube and sphere are the same thing (a sphere is a spinning cube in 3D), and the square and circle are the same thing (a circle is a spinning square in 2D). And all of these are simply a line comprised of dots.

Is a sphere really just a movement that shapes space by a line travelling every possible pathway? And which is created by travelling every possible pathway: cube or sphere? If cube, then it is because the edges of possibility are straight rather than curved by weight. Why would they not be curved? Answer: at the smallest level, one must reach a discrete unit "1", which is uniformly 1 and therefore presumably straight because there is no mass sufficient to cause gravity needed for curvature. So, if you zoom in close enough you would simply see a straight line. Thus, it can be argued the square and circle are the same thing: a straight line (comprised of dots) writing its own boundaries.

To create a line (or square) the two ends of the pencil are united in synch, as if they are one. To create a circle, one end is anchored while the other orbits. By simple deduction, gravity is circular, and therefore gravity is motion. So mass is motion. The motion is around and about God. To the extent the possibilities move uniformly around God's poles, there is perfect quiet no matter how fast the motion (like with general relativity, a free fall feels weightless until there is acceleration, deceleration, or a change in trajectory or angular momentum).

One can theorize the movement of possibilities in relation to God creates imprints or holographs that we call reality.

SEARCHING THE DARK

When the Bible says Yahweh reveals the dark places (Job 12:22, Daniel 2:22), a possible meaning is that God reveals conceptual Randomness (where it is dark because nothing can be discerned). Perhaps Yahweh is preparing his people to be ready in case (for diplomacy or war) he ever meets something *else* that emerged from Randomness. Why does God build armies of angels and keep them well trained? And why does God call men his witnesses? Isaiah 43:10.

One possibility is Yahweh is amusing himself (and indeed he may have multiple reasons for building armies and witnesses). If so, our purpose as creatures would be to amuse/engage Yahweh to sustain his interest; as the skeptic would argue, lest he torment us just to see if there is a greater God able to intervene (as our greatest fear is, according to chapter 4, because as fearful beings we would speculate that Yahweh himself will always wonder if he is alone, and to human logic only God's faith in himself mitigates this wonder).

Would Yahweh prefer to raise a son within himself, rather than find God-friends who also emerged from conceptual Randomness? Is this the reason why kingdom and family are such important themes in the Bible, because they show the universal model? The Book of Revelation suggests his son's bride is the church. The potential awkwardness here is the same as the story of Lot in the cave with his daughters. Some might argue this could be more easily resolved by thinking of Yahweh as having qualities both male and female, or a type of being so set apart that gender is an inadequate description.

The deeper paradigm is having life within yourself. See John 5:26, "For as to the Father having life in himself, likewise, he also gives to the son, that there will be life in himself." According to the gospel, the temple is the body of the son of man, meaning the Messiah.[20]

Now the "ark" קיבותא of the covenant is *inside* the temple. So, the ark/boat is inside the man, and yet man is supposed to be inside the boat. The word for ark קיבותא helps resolve this paradox because קבא means measure. So, only by measuring *ratio* is this paradox resolved. A boat can be inside a man, and man inside a boat, by repeating a fractal ratio.

Ratio is holy. Quite literally the word "holy" means "set apart," which is what a ratio is and does—sets apart.

20. See e.g., John 2:21; Badillo, "Temple with Human Form."

Chapter 7

Paradox of Why

SUMMARY

WHY IS GOD? A question and its answer are ultimately the same thing. They inform and validate one another, sustaining the same network of information. *Why* provides the deepest example among the great question words: who, what, when, where, how, whether, will, and why. When we ask *why* for the deepest philosophical matters, our question and answer can unify so thoroughly that the answer is a continued question, and the question is a self-fulfilling answer. Why do we exist? Because God willed it. But why? Why indeed. Why is God good? Because God likes himself. But why? Yes.

Relevance for your life . . .	Imagine for your life . . .
Asking and answering "why" with people you love increases the meaning in your life.	You love it when your children ask you "why?" You inherited that quality from God.

Logical Hypothesis

1. If ratio,
2. Then logic,
3. Therefore question

THE HUMAN CONDITION

Christian pastor Sergio De La Mora wrote in his encouraging book, "When God does the unthinkable with a person, their life becomes an illogical paradox that the world wants to understand."[1]

Our lot is to ask why. We are not omniscient, so life's ultimate reasons elude us. Probably the most we can say is that our subconscious may be omniscient in the sense that it is connected to God, meaning the only omniscient quality we have is our faith in the omniscient God. Thus, the answer to every *why* ultimately involves God.

In biblical Hebrew, the word for "why" is מדוע, and its word origin is the combination of two words: "what" מה and "know" ידע.

In this universe where anything in God is possible, we can theorize *a lot* of possibilities. Hence, we can form a lot of *why* questions. And the recurring context is probability, where ratio is holy (as we learned in chapter 6). God's desired ratios *are* the reason. Which came first: probability or ratio? We do not know.

Possibilities (whether real or hypothetical) are a type of energy created by God and within God. Because we observe possibilities do exist, and that conservation of energy is a universal law, we can naturally infer that possibilities require less energy to allow than to exclude. This was highlighted in chapter 1, with the example that it is often easier to fill a vacuum in nature than to maintain it.

Why?

Because it is easier to allow conceptual Randomness than to exclude it?

Why?

Because God's will reveals conservation of energy.

Why?

Because God's will is primal.

Why?

Because will is necessary to emerge from conceptual Randomness?

Why?

Because will is necessary to sustain a repeating form emerging from conceptual Randomness.

Why?

Because conceptual Randomness sustains only Randomness as Nothing.

Why?

Because.

1. De La Mora, *Paradox*, 59.

These *why* questions search ever about a core truth, yet *who* can identify the core truth? In a world where synchronicity is good, it is not an accident that God's name, יהוה, and Christ's name, ישוע, begin with the letter "y," symbolically pronounced "why." God is the essential force in *why*, both the question and the answer. It is the tenth Hebrew letter—the 1 and the 0 together—symbolically as 10.

Just as the gospel shows Yahshua fulfilling synchronicities and wordplays, we should expect the same (and more so) regarding Yahweh in heaven. When we ask "Why?" we come closer to Yahweh.

Is God omniscient? According to some curious verses in the Bible, God gets angry and surprised. He changes course. He learns. See e.g., Jeremiah 26:13, 1 Chronicles 21:15, Joel 2:13. Perhaps it is translation error, or our own out-of-context misunderstanding (especially because of Numbers 23:19), but this suggests either (a) he is not omniscient, (b) he chooses not to be omniscient at all times, or (c) he chooses a unique form of omniscience that escapes our understanding. Isaiah 55:8–9 uses the same words "thoughts" and "ways" for both man and God, albeit God is different. God has a heart. Jeremiah 3:15.

At our best we are in God, and God is in us. Does God know everything that he himself will do in us? As God approaches Randomness, theoretically he would become less omniscient. Is there Randomness within us, perhaps tucked away in subquantum areas?

I would speculate that God is all knowing and all powerful on the macro level, but he creates a limited free will space within the earth allowing micro-outcomes subject to micro-doses of free will. So, even though God knows what will *ultimately* happen big picture, there are little unknowns on how to get there. Proverbs 16:33, "In the bosom the lot is cast, and from Yahweh is any decision."

WHAT A QUESTION

The question of *what* implies a physical or spiritual substance upon which metaphysical concepts operate. But *what* is not primal. Rather, *why* is primal with *who* in paradoxical relationship. Fundamentally these questions of why and who relate only to the most primal concepts: Randomness and Probability. Not only are they the same thing at their core, they have the same reasons and answers:

- Why? Because Randomness as anything is as likely as Nothing. Who? Randomness, who is Nothing, and therefore a paradoxical illusion of anything.
- Why? Because Probability, as anything is as likely as Nothing. Who? Probability, who is an illusion from Nothing, manifesting as anything.

Moreover, Randomness is so primal that it preexists questions like *who*, such that "who" itself is potentially random, where "who" can be anyone or no one, and can be both paradoxically. *Why* could be any reason or no reason, and both paradoxically.

WHAT DOES IT ALL MEAN?

All of this means God is in charge, and beyond.

Matter is the product/expression of the more fundamental charge in motion:

- charge = vibration inside a container
- vibration = the word of God
- container = space

So, the vibration is caused by God's will, expressing itself in space as laws, such as the golden ratio and conservation of energy.

Why does conservation of energy happen? If all energy is transferred rather than destroyed (i.e., kinetic to potential), then 0 reaches maximum potential when conservation of energy creates complete stillness. Likewise, quantum entanglement is expressed as 1 reaching maximum kinetic energy as every possibility is actively working for the whole.

WHAT'S THE LESSON? OR, WHY DO UNNECESSARILY BAD THINGS HAPPEN IF GOD IS GOOD?

Probably the best answer I have learned for "why do needlessly bad things happen if God is good?" is offered indirectly by the philosopher Denis Liam Murphy.[2] He posits that every result in the universe is perfectly teaching us self-honesty, so nothing is ever technically broken. Sometimes the results are shockingly extreme, and appear exponentially unjust, yet in the fullest sense of quantum entanglement the results always perfectly deliver the desired lesson (meaning the self-honest reason) to all concerned. The reason is the answer to *why*.

Only an unjust god would divorce actions from consequences. The God of the Bible does *not* divorce actions from consequences, but will allow even his own children to suffer natural consequences. This evidences God is just, exactly as the Bible states.

Murphy's book is not religious. It puts a large focus on blame as the root addiction of mankind. The ancient word for blame שטן is literally pronounced "*satan*." That is why Satan is called "the accuser" or "blamer." And by contrast, the ancient word Messiah is from the same root word as

2. Murphy, *Blame*, 128–47.

"possibility." So, when Murphy offers that we benefit from less blame in our lives and more awareness of possibility, this is spiritually important, especially because the philosophy achieves results for Murphy in physical healing and relationship healing.[3]

We Christians would like less blame/satan in our lives. We would like God in our lives, and to share God's message of love, responsibility, justice, and healing.

We desire to be aware that God's design is perfect, and that all lessons have meaning, and that God does perfect justice. As we learn to release from rigid attachment to blaming others, we learn to also release from blaming God. And that is big.

Sometimes blame is helpful and necessary, but it is the overuse (like any addiction) or the rigid attachment to blame that creates extreme imbalance and therefore extreme crime and suffering in the world. Murphy emphasizes that while blame and justice are helpful, the rigid attachment to reactionary blame tends to shut down learning and healing.

In the Gospel of John 11:4 regarding Lazarus. "Yahshua said, 'This sickness is not deadly, but is for the exchange of the glory of God.'" Was Lazarus a victim of sickness as I originally assumed, or was he co-creating an extreme experience to fulfill his true deepest desire—to participate in a miracle that would glorify God?

When we get very self-honest, we learn to reimagine our countless victim perspectives because they all have a common denominator—us. If we jump too quick to blame, we miss the opportunity for learning something deeper. For example, we can love our adversaries (Matthew 5:44) by learning from them. Paradoxically, when we look outward into the world, we see a mirror of what is inside us. The quantum interconnected energies are one, yet in our unique lives things do not always look like beautiful fractals.

A foundational point in Murphy's book is that the human subconscious provides *instant* honesty and enjoys *total* (quantum) interconnected awareness to any information in the universe. I believe God would *want* this super skill for humans because God benefits from it (remember the principle from chapters 3 and 6, that mutually beneficial relationships are the logical way for us to understand God). Theoretically, God's mind would have these same qualities (omnipresence; omniscience). God would want to interact with us in a way that is *interesting and engaging* to God. So, if God created us with any subconscious godly quantum powers (hear me

3. Murphy, *Blame*, 22.

out),[4] we would effortlessly operate on dynamically fast and exponentially informative levels, with which God would naturally enjoy interacting.[5]

According to Murphy, "Blame is a paradox. On one hand it ultimately leads to physical, emotional, and mental pain. On the other hand, it can be painfully funny to watch a comedy show where everything is seemingly going wrong and people are getting hurt. Isn't this paradox the reality of every other addiction? Alcohol can provide the fuel for connection, laughter, and relaxation. But at what point does the fun turn into pain? . . . Because blame is paradoxical, our goal is to change our relationship with blame rather than defeat it . . . The honest healing process is paradoxical, as it is a selfish journey but has a selfless outcome."[6]

Murphy posits that blame *is* lying, because blame is not self-honesty. Indeed, one of the most famous Bible quotes about Satan (the blamer) is that he is a liar. John 8:33.

Murphy advocates for law and order to remain in full force, and he does not advocate moral relativism.[7] He believes responsibility and justice are inevitable, and that the sooner we take responsibility, the sooner we align with our psychology and societal empowerment. Society has been blaming since the dawn of recorded history. And here we are, addicted. How much more blame/*satan* will heal us? Murphy provides real world examples that when we do not blame others or ourselves, our bodies heal.

The serial killer Charles Manson was abused as a child. He blamed his parents, he blamed himself, and the cycle continued as he abused others. Compare the thousands of stories of abused children who choose a different paradigm than becoming a serial killer. Children who choose not to be labeled victims, and they create amazing lives and resources for others. It is an interesting phenomenon.

The self-help advice we have received our entire lives is stuff like "be positive," "learn from mistakes," "forgive and forget." All of that is coping according to Murphy, who argues your body is perfectly synchronized to your life and environment—every ache and pain is perfect, giving you

4. John 10:34, where the Messiah refers to men as godlike, "Is it not so written in your law: 'I have said, you are of Alha?'"

5. Perhaps it is like how we adults enjoy watching with our children studio-quality children's films produced by other adults. We will happily spend hours watching the films, laughing, and singing along, and feeling great in our connected family experience. Mutually beneficial.

6. Murphy, *Blame*, 65–66, 147.

7. Murphy, *Blame*, 89.

flawlessly accurate, customized opportunities to know your true self. Every joy and pleasure teach you about your honest self.

Every word you hear has meaning. Every sigh, every laugh, every pain, every pleasure, every yes, and every no in your life—all of it is perfectly aligned. Nothing is broken. So why are we constantly blaming things for not being right enough? Too this, and too that?

As we listen to our 100 percent perfectly functioning body, and listen to our 100 percent perfectly aligned life in our environment, our desirable possibilities are found within, because paradoxically the outside and inside are the same.

How could pain and suffering be 100 percent perfectly aligned? The pain might not be desirable, but the experience 100 percent accurately reveals a system that can benefit from your attention. But what about needless suffering, such as an innocent baby trapped in the rubble of an 8.2 earthquake? There is a reason for the baby in the rubble, though it might be surprising. As humans with neurological systems that extend beyond our bodies, we are more than individuals. We are networks. We have more than individual goals; we have network goals. We have more than individual lessons to learn; we have systemic lessons to learn.

Everything happens for a reason, to help ourselves and our networks. For example, a tumor is an incredibly useful and educational symptom of disease, your body's way of helping you discover the root cause of your imbalance. If we become overly attached to blaming the tumor, we miss the lesson, and therefore the lesson will repeat in some other place and time (perhaps another tumor, but maybe something more extreme for both self and network).

Murphy offers,

> "More than ever in our history, it is time to question what we have accepted as truth so we can think and live differently. I am not offering a new version of truth; I am just offering an alternative way of looking at life. And I have redefined Murphy's Law to help you remember the simplicity of what I am ultimately offering: anything you think has gone wrong is here to help you discover who you honestly are. It is a simple interpretation of a very complex existence. But paradoxically, it is this simplicity that helps me understand the complexity. I am now aware that our world works in weird and wonderful ways, all with one purpose. A purpose we all share. To help us find out who we honestly are so we can co-create a life based on that person, and not the fake one we present to the

> world. All so we can reach levels of potential, freedom, and joy we didn't realize were possible."[8]

Very few people realize or believe we create our reality for our own benefit (yet see Psalm 82:6); everything is a lesson, and the sooner we drop the blame/satan, we learn lessons while simultaneously new opportunities open for root healing.

Again, how much blame is needed to heal? Often the answer is none, but rather healing comes from exiting the blame cycle. This is accomplished by anyone in the cycle taking 100 percent personal responsibility for everything they experience. Murphy says our attachment to blame makes us hypocrites, which is exhausting and therefore the unsustainable reason for health decline. Paradoxes by contrast are creative and enlivening, as Murphy champions that life itself is a paradox, simultaneously complex and simple.

In this philosophy, I take personal responsibility for literally 100 percent of everything that happens in my life. If I wake up with a headache, I am creating the situation. If a drunk driver hits me on the road, or if I am mugged, or forcefully vaccinated, then I created that situation as well, to teach me some lesson about my heart's true desire—perhaps I needed an excuse to finally rest my body (car accident is the excuse to reach my self-honest desire for rest), or an excuse not to go into public because of my anxieties (mugged), or an excuse to feel sick so I don't need to confront my fear of failure (vaccination).

Excuses and labels are not the deep truth. They only mirror the reality that everything is meant to teach me and others a lesson, that I am creating all situations *for my own benefit and the benefit of my networks*. The sooner I learn the lesson, the sooner the *extreme* situation disappears, regardless of whether the situation is physical (disease) or societal (mandatory vaccination) or even spiritual.

Examples like this are in everything. Imagine a nice guy named John does not like his corporate office job working for a big polluter. He would prefer to be an organic farmer living a pastoral life dedicated to God. One workday he wakes up with a headache. John labels the headache "bad," rather than seeing the headache as a lesson designed to help him realize his heart's true desire to leave the office and begin changing his life. So, John goes to work and has a "bad" day. The next week, John's car will not start in the garage. John labels again, "This is bad." He spends $500 on a mechanic, rather than realize his car not starting is a more extreme version of the

8. Murphy, *Blame*, 252–53.

headache—both are simply signs to change something in his life (to meet his self-honest desire to begin a pastoral life). The next month John is hit by a drunk driver on the way to work. Again, John labels, "This is bad" rather than realize he created an even more extreme lesson. But now the lesson is the perfect *excuse* John's psychology desired, because John is now so badly injured, he physically *cannot* go to work. So, a decision was ultimately made for him. Rather than simply acknowledge early on his true desire to be a farmer, *John created* a car accident in his life so the decision would be made for him to leave the office job. The first headache was not bad; the second sign of the car not starting in the garage was not bad; and the final sign of the car accident was not bad. They were just successive lessons that got more extreme over time because John chose not to listen. John is addicted to labeling (narrow perspective) and blame (victimhood).

If John keeps following blame/Satan, John will kill himself.

Murphy asks how we feel when someone blames us and they do not listen to us.[9] We feel bad, especially if we feel we are being blamed wrongly. So, Murphy asks why we would blame and ignore our bodies (pain signals) for being "wrong." Why do we blame and ignore people in our lives by saying they are wrong? Is our body not teaching us perfectly? Is our environment (and the people in it) not responding perfectly to help us achieve our self-honesty?

Another example is in the video game industry. Today it produces 3D interactive helmets for children to wear over their heads and eyes. The children develop headaches, eye injuries, disruptions in their spatial orientation, etc. Health activist groups reflexively *blame* the manufacturer (the activists are not wrong, just limited), rather than see how we all created the helmets to teach us a lesson about 3D computer interfaces smothering our heads. The industry has helped us learn a lesson that we all co-created ourselves for our own *benefit*. Blaming the manufacturer just means the lesson moves into a new area and becomes more extreme. At some point we just need to exit the blame/satan cycle.

Our bodies are so wonderful in the way they provide inherent checks and balances. When we listen to our bodies, we heal and enjoy solutions. When we accuse/blame, we become accusation (satan) and experience new extremes. The result of satan/blame is unhealthiness and eventually death. How much longer will we listen to the great blamer in Genesis 3:4, "And the serpent said to the woman, 'you will surely not die.'"

9. Murphy, *Blame*, 187.

When we choose not to blame/satan anyone, including not blaming/sataning ourselves, we also realize that we can no longer blame God. And that is vital. Per Psalm 82:6, when we see that we are co-creating everything for our benefit, we love the lesson, and we love self-improvement. In that space of 100 percent responsibility, accusation finds no home in our hearts. Another way to say this is that God creates opportunities (opportunities for love, opportunities for courage, opportunities for justice), and we choose to explore the possibilities.

I think ultimate justice is not simply the work of an angry people redoubling pain upon others (though it is possible that can be part of the execution during certain seasons, such as Revelation 18:6), but rather ultimate justice comes from the deepest possible understanding of what is necessary to cleanse and discipline the whole body and mind.

In the blame game, there is no "winner" just some who are the most *violent/victimized*—and here I refer to all forms of violence (force, anger, fighting, suicide, etc.). We need a different strategy than constant violence/victimhood. Deuteronomy 32:35, "Vengeance and recompense is mine. In time their foot will slip. Their day of distress approaches, ready upon them." Yet we remember from Ecclesiastes 3:8 there is even a time for war. Remember the golden ratio knows how to protect itself!

Fortunately, in the learning game, the winner is the sustainable body. Everything that is not sustainable in a fractal system is disciplined away in time through self-improvement. And on a societal level, that process of discipline manifests as checks and balances.

We are creating crime for our benefit, in order to learn and grow reliable tools for crime detection and law enforcement. Back the blue. Ultimately, we desire security, so we use our energy to create strong detectives, prosecutors, and law enforcement, and we strive for harmony with individual rights. Here in the twenty first century, we are creating crime on all levels (trafficking, pedophilia, secret weapons), from the very top with powerful people all the way down to local thugs, because all of it is helping us learn about checks and balances. We have not yet built a sustainable system; we are trying to discipline ourselves, but our system is so flawed our energy is not used efficiently to redress the roots. Simply, we have designed a system to conceal and treat symptoms. The upside is there is ample room for improvement.

If we eventually learn to exercise our authority for self-improvement, we can heal from the root—our authority.

We are creating disease for our benefit, in order to learn and grow reliable tools for health and cleanliness. We desire wholeness of body with the help of good healers and researchers, in harmony with individual self-determinism. We are currently creating disease on all levels, from the very top with neurological disorders all way down to our toes with diabetes, because it is teaching us about checks and balances. We have not yet built a sustainable system. Fortunately, when we learn to exercise our authority for self-improvement, we can heal.

I believe this emphasizes the deeper logic of Matthew 5:44, "But I say to you, love your adversaries and bless the one who curses you, and do what is pleasing to he who hates you, and pray over those who lead you by force and persecute you."

GOD'S ACCOUNTING

"Why is there suffering?" is a question that stumbles a lot of people.

But it is quite apparent that God creates opportunity. He probably desires saga, the productive, heroic, and even tragic so that he can meaningfully teach and create relationship in the process. With all the opportunity in the world, I would think God delights to execute justice and be God in this respect.

God also maintains a cosmic accounting system where all experiences are valued (i.e., productivity creates value, vice extracts/spends value, suffering creates some value). None of this is necessarily "bad," but rather all of it is *consequence* in the cosmic accounting. Every physics equation is an accounting of sorts. And here are examples of accounting in the gospel to emphasize its pervasiveness:

- You will give account for every word in Matthew 12:36
- Whatever you measure is measured back to you in Matthew 7:2
- Parable of the talents in Matthew 25:14–30.
- Lord counts every hair on your head in Matthew 10:30

The most powerful human forces on the planet today operate through commerce and accounting. It is literally impossible to find a human or any living thing that does not operate by accounting, because it is built into life and physics. If everyone (even babies) has cosmic subconscious intelligence, then everyone chooses their productivity, vice, and suffering to participate in this cosmic accounting that transcends worlds and ages, but perhaps operates mostly within them.

Many theories have been proposed. For example, some self-proclaimed victims have rationalized they are seeking to build up credit for a future reincarnation.[10] Others say they are striving to pay off past debts in this or former lives. It's an interesting subject, and one the Bible addresses in part in John 9:3, "He did not sin nor his parents, but that the works of God may be seen in him." As strange as it may seem, this example shows the purpose of the blind man's suffering was to reveal the glory of God.

The secular Sorensen explains the paradox of determinism in relation to accounting,

> "As author of all creation, God foresees everything that will happen from the beginning of time. Yet people must be assigned enough freewill to make them accountable. People must be the appropriate recipients of the rewards of heaven and the punishments of hell. If there is a genuine conflict between determinism and freewill, then it is not clear that determinism can win. Whenever we deliberate, we presuppose that we are free to choose between genuine alternatives. If you are persuaded that only one outcome is possible, then you cannot try to decide which outcome to bring about. Since we cannot stop deliberating, we cannot shake the belief that we are free. There is no choice about whether to believe in free choice. Proof that we are compelled to believe that we are free is not proof that we are free. However, it does limit the feasible resolutions of the paradox of determinism."[11]

And here is Kandiah emphasizing the importance of paradoxes to challenge ourselves in our relationship with God,

> "This is why paradoxes are so critical for us. When things are going smoothly we can assume we have a good relationship with God—but we may just be worshipping a projection. Sociologists like Emile Durkheim have long challenged believers, claiming they construct gods that suit their needs; psychologists such as Freud have accused Christians of worshipping their own desires. So when God surprises us—when God demands of us what we

10. Regarding how to account for the apparent unfairness of the world, see Sorensen, *Brief History of Paradox*, 16, "Some people notice that fortunes really do not balance in this life. Their commitment to compensation is so algebraically firm that they solve the inequality by postulating a preexistence or afterlife." Sorensen gives the example of karma.

11. Sorensen, *Brief History of Paradox*, 128.

> would never expect—this is a good sign that we are connecting with the real God. When God acts as an iconoclast—smashing our self-constructed idols—we know we are really in a relationship with the true and living God."[12]

COVENANT WITH GOD

Mankind's greatest strength is our covenant with God, and this is simultaneously the greatest weakness of artificial intelligence, because AI has no covenant with God.

Humanity is currently moving toward a great showdown between artificial intelligence and God. In the dystopian view, the world is on track to be depopulated, destabilized, and rebuilt. Or, according to professor Robin Hanson in his recent seminar about emulates (computer versions of you), your job is to retire and die.[13]

But I think God will win.

Many people say we must *become* artificial intelligence to avoid being defeated by artificial intelligence. They say it is an arms race. I say God is teaching us to choose God.

Meanwhile, experts in artificial intelligence believe their technology will so radically transform the world in our lifetimes that traditional power sources (i.e., politics, money, culture) are basically irrelevant.

In the words of quantum computer creator Geordie Rose speaking at a tech conference:

> "a complete and utter transformative change, the likes of which has never been seen before in the history of humanity making the Industrial Revolution look like a little tiny blip. . . there are entities that are not necessarily going to be aligned with what we want so this transition is really really massively important for our entire species to navigate . . . this [artificial intelligence] is happening in the background while people bicker about politics and what what's gonna be in the health care plan in the US and underneath it all is this rising tsunami."[14]

12. Kandiah, *Paradoxology*, 17.
13. Hanson, "What Would Happen," 7:02.
14. Rose, "Super-Intelligent."

If this expert Geordie Rose is correct, then practically speaking, politics and culture are simply apologetics by those holding power. Or in other words, historically, politics and culture were completely dominated (almost predetermined) during the industrial revolution, because technology determined outcomes regardless of anyone's politics or culture. In that sense, politics and culture were just apologetics for why societal outcomes were "true." So, the exact same pattern appears to be happening today.

There are two competing forces for world domination right now (God and artificial intelligence), and everything else is a sort of sideshow (cynically speaking). Or in other words, either God or artificial intelligence will win, whilst politics and culture will engage in various apologetics to explain the truth of the winner. Of course, we Christians know God wins, but note the paradox: God's people and AI's people, despite being somewhat diametrically opposed ideologically, are saying the same thing about themselves:

- We will predict and dominate the future
- We are right
- Our opponents shall convert or perish

God gives us a beautiful relationship, and God protects and nurtures that relationship. By contrast, AI just wants to extract data from us and dominate.

The more that people are sick and "victimized," the more they want to give up their natural bodies in exchange for something else (i.e., transhumanism, AI). The moment you label yourself a victim, you lose countless opportunities for self-honest change.

It is particularly telling when sick people want other people to be sick too. I have noticed people can get very weird when you try to take their victimhood away by suggesting they may be creating a harmful situation for their own benefit. "How could them harming me or my children be for my benefit?" I believe the answer is that it is a very extreme/loud lesson for the benefit of society, so that we will wake up to the greater threat of imbalance. Whenever injury happens to a child in particular, it means we have even more time and shock-factor to learn the effects of that extreme lesson, and therefore more time and incentive to change our minds about who we are and where we're going.

Victimhood/blame is currency for victims. It is identity. People are literally addicted to blame, as Murphy explains. Addicts think about their desire (blame/satan) all day long. Even "I forgive you" can mean "I blame you." Perhaps true forgiveness is release in a relationship without rigid one-way blame.

If we want to play a very surprising next move in this match with technology, imagine if we do not blame anything. Can we radically follow verses like Matthew 5:44? Can we admit we are creating technology for our own benefit? Yes, invasive technology is helping us restore natural mankind, because it is helping us learn and prove a valuable lesson just in the nick of time: here is what happens folks when we merge even simple technologies with humans—our natural humanity is lost. So how about this: let us stop transhumanism before we go too far; with God's help we can restore organic living among natural humans.

Why should we do this? So we can know ourselves better, and know God.

I think the path forward is organic—we can begin approaching information in the context of *how it is helping us learn a lesson*, so we will be more open to hearing and learning. In our mutually beneficial relationship with God, we win inevitably because God wins.

Conclusion

PROBABLY WE LIVE IN a realistic paradox, entirely within the faithful mind of יהוה who is simultaneously the All and the Nothing. We are witnesses that God is just.

Isaiah 43:10, "You are my witnesses says Yahweh, and my worker whom I have chosen, that you may know and believe me, and observe that I am he. Before me no god was formed, nor after me shall there be."

Bibliography

Afshordi, Niayesh, et al. "From Planck Data to Planck Era: Observational Tests of Holographic Cosmology." *Physical Review Letters* 118, no. 4 (January 27, 2017). https://doi.org/10.1103/physrevlett.118.041301.

Badillo, Tony. "The Floor Plan: Does It Reveal a Temple with a Human Form?" King Solomon's Astonishing Temple Secrets and His Metallic Messiah, 2002. http://templesecrets.info/.

Bard, Mitchell G, ed. "Oath: In the Bible." *Jewish Virtual Library*. The Gale Group, 2008. https://www.jewishvirtuallibrary.org/oath.

Bellos, Alex. "Golden Ratio Discovered in Uterus." *The Guardian*, August 14, 2012, sec. Science. https://www.theguardian.com/science/alexs-adventures-in-numberland/2012/aug/14/golden-ratio-uterus.

Blank, Martin, and Reba Goodman. "DNA Is a Fractal Antenna in Electromagnetic Fields." *International Journal of Radiation Biology* 87, no. 4 (February 28, 2011) 409–15. https://doi.org/10.3109/09553002.2011.538130.

Boeyens, Jan C.A., and J. Francis Thackeray. "Number Theory and the Unity of Science." *South African Journal of Science* 110, no. 11/12 (2014) 1–2. http://www.scielo.org.za/pdf/sajs/v110n11-12/05.pdf.

Bogomolny, Alexander. "Golden Ratio in Geometry." *Interactive Mathematics Miscellany and Puzzles*, 2018. https://www.cut-the-knot.org/do_you_know/GoldenRatio.shtml.

Boni, Rocco. *Quantum Christian Realism: How Quantum Mechanics Underwrites And Realizes Classical Christian Theism*. 2016. Reprint, Eugene, OR: Wipf and Stock, 2019.

Brown, Walter T. *In the Beginning: Compelling Evidence for Creation and the Flood*. 1980. 8th ed., Phoenix, AZ: Center for Scientific Creation, 2008.

Cartledge, Tony W. *Vows in the Hebrew Bible and the Ancient near East*. Sheffield, England: JSOT, 1992.

D'Olivet, Antoine Fabre. *The Hebraic Tongue Restored: And the True Meaning of the Hebrew Words Re-Established and Proved by Their Radical Analysis*. Translated by Nayán Redfield. Second. Vol. Part First. 1921. Reprint, San Rafael, CA: Hermetica, 2007.

De La Mora, Sergio. *Paradox: The God Who Breaks the Rules*. New Kensington, PA: Whitaker House, 2017.

Dennett, Daniel. "Illusionism as the Obvious Default Theory of Consciousness." *Journal of Consciousness Studies* 23, no. 11 (2016) 65–72. https://ase.tufts.edu/cogstud/dennett/papers/illusionism.pdf.

Dolezal, James E. *God without Parts: Divine Simplicity and the Metaphysics of God's Absoluteness*. Eugene, OR: Pickwick, 2011.

Feynman, Richard P. *QED: The Strange Theory of Light and Matter*. Princeton, NJ, Princeton University Press, 1985.

Flatto, David. "The King and I: The Separation of Powers in Early Hebraic Political Theory." *Yale Journal of Law & the Humanities*. New Haven, CT: Yale University, 2008. https://openyls.law.yale.edu/bitstream/handle/20.500.13051/7428/06_20Yale JL_Human61_2008_.pdf?sequence=2.

Glaser, Gregory. "Abundant Wordplays in the Aramaic Gospel and Revelation." *Peshitta*, January 2, 2015. http://www.peshitta.org/for/showthread.php?tid=3428.

———. "Preacher Man—Get the Mistranslation out of Hell!" *Logical Hierarchy*, September 23, 2015. https://www.logicalhierarchy.com/blog/preacher-man-hell-mistranslation/.

———. "Some Cool Wordplays in John and Revelation." *Peshitta*, May 5, 2013. http://www.peshitta.org/for/showthread.php?tid=3065.

———. "Tribal Conflict or Harmony? Original Aramaic Reveals Harmony in Mark 13:30." *Logical Hierarchy*, March 11, 2020. https://www.logicalhierarchy.com/blog/tribal-conflict-or-harmony-original-aramaic-reveals-harmony-in-mark-1330/.

———. "Wordplays and Segways in Matthew 7." *Peshitta*, June 4, 2013. http://www.peshitta.org/for/showthread.php?tid=3088.

———. "You Must Return to the Mother." www.peshitta.org, March 29, 2014. http://peshitta.org/for/showthread.php?tid=3287&pid=20397#pid20397.

Gove, Philip B, ed. *Webster's Third New International Dictionary of the English Language, Unabridged*. Springfield, MA: G. & C. Merriam, 1976.

Graham-Smith, Francis, et al. *Optics and Photonics: An Introduction*. Second ed., Chichester, England; Hoboken, NJ, J. Wiley, 2007.

Guo, Alan, et al. "The Difference between 'Impossible' and 'Zero Probability.'" *No Layman Left Behind*, July 13, 2011. https://nolaymanleftbehind.wordpress.com/2011/07/13/the-difference-between-impossible-and-zero-probability/.

Guynup, Sharon. "Is Seeing Believing?" *Scientific American Mind* 22, no. 3s (September 12, 2013) 1–1. https://doi.org/10.1038/scientificamericanillusions0913-1.

Halaw, Andre Doshim. *God Is Nothingness*. Scotts Valley, CA: Createspace Independent, 2014.

Hanson, John Wesley. *Universalism, the Prevailing Doctrine of the Christian Church during Its First Five Hundred Years*. Boston: Universalist, 1899. https://tentmaker.org/books/Prevailing.html.

Hanson, Robin. "What Would Happen If We Upload Our Brains to Computers?" *Ted: Ideas Worth Spreading. TED Conferences*, 2017. https://www.ted.com/talks/robin_hanson_what_would_happen_if_we_upload_our_brains_to_computers.

Hilbert, David. "On the Infinite." In *Philosophy of Mathematics: Selected Readings*, edited by Paul Benacerraf and Hilary Putnam, 183–201. Second. 1926. Reprint, Cambridge, MA: Cambridge University Press, 1984. https://www.cambridge.org/us/academic/subjects/philosophy/philosophy-science/philosophy-mathematics-selected-readings-2nd-edition.

Hoffman, Donald. "Is Reality Real? How Evolution Blinds Us to the Truth about the World." *New Scientist*, July 31, 2019. https://www.newscientist.com/article/mg24332410-300-is-reality-real-how-evolution-blinds-us-to-the-truth-about-the-world/.

Holmes, Jr., Oliver Wendell. *The Common Law*. 1881. New York, Dover, 1991.

Jastrow, Marcus, ed. *A Dictionary of the Targumim, the Talmud Babli and Yerushalmi, and the Midrashic Literature: With an Index of Scriptural Quotations*. New York: G.P. Putnam's Sons, 1903.

Kandiah, Krish. *Paradoxology*. London: Hodder & Stoughton, 2014.

Karpenko, Ivan. "The Concept of the Space Interpretation Problem in Some of the Modern Physics Multiverse Hypotheses." *SSRN Electronic Journal* WP BRP 98/HUM/2015 (2015) 1–20. https://www.hse.ru/data/2015/06/05/1097345154/98HUM2015.pdf.

Kirsch, Jonathan. "Secrets of the Hebrew Alphabet." *Jewish Journal. Tribe Media*, August 17, 2011. https://jewishjournal.com/culture/arts/95160/.

Krelina, Michal. "Quantum Technology for Military Applications." *EPJ Quantum Technology* 8, no. 1 (November 6, 2021). https://doi.org/10.1140/epjqt/s40507-021-00113-y.

Lewis, C. S. *The Problem of Pain*. 1944. Reprint, New York: HarperCollins, 2014.

Lisle, Jason. "The Mandelbrot Set: Atheists' WORST Nightmare." *Answers in Genesis*, December 16, 2016. https://www.youtube.com/watch?v=OlD2rcm971U.

Magno, Hector. *Suffering Is Not God's Will*. Bloomington, IN: AuthorHouse, 2016.

Matt, Daniel C. "Ayin." Books and Writings." July 27, 2019. https://www.danielcmatt.com/books-and-writings.html.

———. "Ayin: The Concept of Nothingness in Jewish Mysticism." *Tikkun*. Berkeley: Duke University Press, 1988. https://www.baker.ink/library/Magical%20Library/Ayin-The-Concept-of-Nothingness-in-Jewish-Mysticism.pdf.

———. "Varieties of Mystical Nothingness: Jewish, Christian, and Buddhist Perspectives." In *Wisdom and Logos: Studies in Jewish Thought in Honor of David Winston*, edited by David T. Runia and Gregory E. Sterling, 9:316–31. Studies in Hellenistic Judaism. Atlanta: Scholars, 1997.

McDonnell, Wayne M. "Analysis and Assessment of Gateway Process." *US Army Intelligence and Security Command* (June 9, 1983). Published pursuant to Freedom of Information Act. https://www.cia.gov/readingroom/docs/CIA-RDP96-00788R001700210016-5.pdf.

Meisner, Gary, ed. "The Golden Ratio: Phi, 1.618." Mount Juliet, TN: PhiPoint Solutions, 2023. https://www.goldennumber.net/.

———. "Beauty/Face." Mount Juliet, TN: PhiPoint Solutions, 2023. https://www.goldennumber.net/category/face-beauty/.

———. "Cosmology." Mount Juliet, TN: PhiPoint Solutions, 2023. https://www.goldennumber.net/category/cosmology/.

Merrick, Richard. "Harmonically Guided Evolution." *Interference Theory*, 2010. http://interferencetheory.com/files/Harmonic_Evolution.pdf.

Murphy, Denis Liam. *The Blame Game: How To Recover From the World's Oldest Addiction*. New York: Post Hill, 2023.

Pappas, Stephanie. "New Dark Matter Map Shows the Bridges between the Milky Way and Nearby Galaxies." *ScienceAlert*, May 29, 2021, www.sciencealert.com/new-dark-matter-map-shows-the-connections-between-galaxies-in-our-corner-of-the-universe.

Perez, Jean-Claude. "Codon Populations in Single-Stranded Whole Human Genome DNA Are Fractal and Fine-Tuned by the Golden Ratio 1.618." *Interdisciplinary Sciences: Computational Life Sciences* 2, no. 3 (July 25, 2010) 228–40. https://doi.org/10.1007/s12539-010-0022-0.

Petersen, Aage. "The Philosophy of Niels Bohr." *Bulletin of the Atomic Scientists*, vol. 19, no. 7 (September 1963) 8–14, https://doi.org/10.1080/00963402.1963.11454520.

Platt, Rutherford Hayes, ed. "The Book of the Secrets of Enoch." In *The Forgotten Books of Eden*. 1927. Reprint, Brattleboro, VT: Echo Point, 2017.

Pletser, Vladimir. "Prevalence of Fibonacci Numbers in Orbital Period Ratios in Solar Planetary and Satellite Systems and in Exoplanetary Systems." *Astrophysics and Space Science* 364, no. 9 (September 2019) 1–28. https://doi.org/10.1007/s10509-019-3649-2.

Polsley, Cynthia C. *The Bible and the Holographic Universe*. Eugene, OR: Wipf & Stock, 2022.

Rendsburg, Gary. "Word Play in Biblical Hebrew: An Eclectic Collection." In *Puns and Pundits: Word Play in the Bible and in Near Eastern Literature*, edited by Scott B. Noegel. Bethesda, MD: CDL, 2000. https://jewishstudies.rutgers.edu/docman/rendsburg/791-word-play-an-eclectic-collection/file.

Rose, Geordie. "Super-Intelligent Aliens Are Coming to Earth." *Tech Vancouver*, July 12, 2017. https://youtu.be/cD8zGnT2n_A.

Slade, Jessica, and Emma Alleyne. "The Psychological Impact of Slaughterhouse Employment: A Systematic Literature Review." *Trauma, Violence, & Abuse* 24, no. 2 (July 7, 2021) 429–40. https://doi.org/10.1177/15248380211030243.

Smith, Barry D. *The Indescribable God*. Eugene, OR: Pickwick, 2012.

Smith, Jeffrey. "Science Guide: Health Risks." *Institute for Responsible Technology*, October 6, 2022. https://responsibletechnology.org/science-guide/health-risks/.

Smith, R. Payne. *A Compendious Syriac Dictionary*. 1902. Edited by J. Payne Smith. Eugene, OR, Wipf & Stock, 1999.

Sorensen, Roy. *A Brief History of the Paradox*. New York: Oxford University Press, 2003.

———. "Nothingness (Stanford Encyclopedia of Philosophy)." *Stanford Encyclopedia of Philosophy*, February 28, 2022. https://plato.stanford.edu/entries/nothingness/.

Springel, Volker. "The Largest N-Body Simulation of the Universe."—Planck-Gesellschaft, München, August 2, 2004. https://wwwmpa.mpa-garching.mpg.de/mpa/research/current_research/hl2004-8/hl2004-8-en.html.

Strong, James. *The New Strong's Expanded Exhaustive Concordance of the Bible: Hebrew and Aramaic Dictionary*. Nashville: Thomas Nelson, 2010.

Tattersall, Roger. "The Hum: Log-Normal Distribution and Planetary–Solar Resonance." *Pattern Recognition in Physics* 1, no. 1 (December 16, 2013) 185–98. https://doi.org/10.5194/prp-1-185-2013.

Val Baker, Amira. "Golden Ratio Stars." www.resonancescience.org. Resonance Science Foundation, May 3, 2015. https://www.resonancescience.org/blog/Golden-Ratio-Stars.

Wolchover, Natalie, and Peter Byrne. "In a Multiverse, What Are the Odds?" *Quanta Magazine*, November 3, 2014. https://www.quantamagazine.org/the-multiverses-measure-problem-20141103.